American Institute of Chemical Engineers

Topical Conference on Path Forward to Hydrogen 2005

Held at the 2005 AIChE Spring National Meeting

April 10-14, 2005
Atlanta, Georgia, USA

Printed from e-media with permission by:

Curran Associates, Inc.
57 Morehouse Lane
Red Hook, NY 12571
www.proceedings.com

ISBN: 978-1-60423-470-1

Some format issues inherent in the e-media version may also appear in this print version.

TD - Path Forward to a Hydrogen Economy

72a Nuclear Hydrogen from a Utility Perspective

Dan Keuter, Steve Melancon

Nuclear Hydrogen from a Utility Perspective

72b U.S. Nuclear Hydrogen Initiative

John F Kotek

U.S. Nuclear Hydrogen Initiative

72c Directions in US Nuclear Hydrogen Production

John H Kolts

Nuclear Hydrogen Production

72d Matching Hydrogen Production Needs with Nuclear Reactor Capabilities: The Advanced High-temperature Reactor

Charles Forsberg

Hydrogen can be produced by thermochemical processes (heat plus water yields hydrogen and oxygen) and by high-temperature electrolysis (heat plus water plus electricity yields hydrogen and oxygen). The high-temperature heat can be provided by nuclear reactors; however, there are several constraints: (1) the cost of delivered heat must be low, (2) for safety, the hydrogen production plant must be separated from the nuclear reactor, and (3) the delivered heat as a function of temperature must match hydrogen production needs. To meet these requirements, a new reactor concept is being developed, the Advanced High-Temperature Reactor (AHTR). The AHTR combines four existing technologies in a new way: (1) coated-particle high-temperature nuclear fuel (similar to that used in gas-cooled reactors), (2) molten-salt reactor coolant (atmospheric boiling points far above operating temperatures), (3) passive safety systems from liquid-metal-cooled reactors, and (4) molten-salt intermediate heat-transport systems to transfer heat from the reactor to the hydrogen production plant.

This combination of technologies is chosen to match nuclear reactor output to hydrogen production requirements. It allows plant sizes of 2400 to 4000 MW(t) to provide economics-of-scale for the hydrogen production. The reactor and heat transport system use molten salts, the technology historically used in the chemical industry to efficiently transport high-temperature heat and isolate energy sources from chemical process units. The high heat capacity of molten salts minimizes the size of pipes and heat exchangers. The high-temperature heat is delivered at a nearly uniform high temperature to the hydrogen production unit to assure that heat from the nuclear reactor will meet process energy requirements. The characteristics of the nuclear reactor, heat transport system, and coupling to the hydrogen production plants are described.

72e The H2-MHR: A Generation IV Concept for Hydrogen Production

Arkal Shenoy, Ken Schultz, Lloyd C Brown, Matt Richards

The U.S. and other nations are sponsoring research and technology development programs aimed at using nuclear power to produce hydrogen safely, economically, and without emissions of (or requiring sequestration of) greenhouse gases. Because of its ability to produce high-temperature helium, the Generation IV reactor technology receiving the most interest for hydrogen production is a modular, passively-safe version of the high-temperature, helium-cooled reactor, referred to as the Modular Helium Reactor (MHR). Two hydrogen-production technologies have emerged as leading candidates for coupling to the MHR: (1) thermochemical water splitting using the sulfur-iodine (SI) process and (2) high-temperature electrolysis (HTE). In this paper, we provide an update on conceptual designs being developed for coupling the MHR to the SI process and HTE. These concepts are referred to as the SI-based H2-MHR and the HTE-based H2-MHR, respectively. Topics of discussion include modifications to the MHR nuclear and thermal hydraulic design for higher-temperature operation, intermediate heat exchanger (IHX) design concepts for transferring high-temperature process heat to the hydrogen production plants, SI-based and HTE-based H2-MHR flow sheets and plant configurations, economic evaluations, and licensing considerations. For high-efficiency hydrogen production, the reactor coolant outlet temperature is expected to be in the range 900 C to 1000 C. Preliminary safety analyses show it should be possible to operate the MHR at higher temperatures without compromising its passive safety characteristics. However, at these temperatures, design of a full-scale IHX (for an SI-based H2-MHR) may prove to be one of the more challenging issues for interfacing the MHR with the hydrogen production plant. In addition, pilot-scale demonstration of the SI process at prototypical temperatures and pressures is required before an SI-based H2-MHR can be deployed. Prerequisites for deployment of an HTE-based plant include demonstration of the Brayton-cycle power conversion system (for supplying electricity to the electrolyzer) and further development of solid-oxide electrolyzer (SOE) technology to improve reliability and reduce unit costs. Although large uncertainties remain at this stage of the design, preliminary economic evaluations show Nth-of-a-kind H2-MHR plants should be able to produce hydrogen economically, and further development of these concepts is warranted.

73a Plant Design and Economic Analysis of a Nuclear Hydrogen Production System

William Summers

Hydrogen production via water-splitting using the heat from a high-temperature nuclear reactor could be an important component of a comprehensive hydrogen economy. Such a system offers high-conversion efficiency, produces no greenhouse gas emissions, has a high degree of siting flexibility, and is capable of very large hydrogen output. During the transition period to a hydrogen economy, nuclear hydrogen plants could be utiized to replace natural gas for large industrial hydrogen users, such as ammonia plants or oil refineries. This paper discusses the design aspects of a nuclear hydrogen plant using a Next Generation nuclear reactor combined with a thermochemical water-slitting process. Plant performance, infrastucture issues, and capital cost estimates are presented. A detailed economic assessment comparing the cost of nuclear hydrogen with hydrogen from conventional sources is discussed, including the computed breakeven price of natural gas where nuclear hydrogen becomes the economic choice.

Molten Salt / Helium Comparison for Intermediate Heat Exchange Loop

*Keith D. Task**
*Edward J. Lahoda***
Lauren Paoletti

Westinghouse Science and Technology Department
1344 Beulah Rd.
Pittsburgh, PA 15235-5083

**Lead Author: 412-256-1260; Taskkd@westinghouse.com*
***Corresponding Author: 412-256-2238; Lahodaej@westinghouse.com*

Abstract- A comparison was made between the possible heat transfer fluids to be used in the intermediate loop that interfaces the sulfur cycle with the Next Generation Nuclear Plant (NGNP) for hydrogen production. From this comparison, an optimal heat transfer fluid was chosen, and design requirements of the loop were set. It has been found that molten salts are preferred to helium for several reasons, not least being reduction in pumping power and heat loss. From the possible molten salts to be used, 50%LiF-50%BeF_2 seems to have the best thermal properties. This salt has not been widely studied, however, and more tests should be run to determine its potential. The salt will be toxic because of its beryllium content, and will probably not be suitable for use because of health issues. Because of this, other safer salts should be investigated with similar thermodynamic and transport properties as 50%LiF-50%BeF_2. Other areas of possible concern include compatibility of the salt with the piping material and with other chemicals in the sulfur process.

1. Introduction

The nuclear industry has the capability of helping to meet the world's increasing demand for hydrogen. The sulfur family of processes utilizes the thermal decomposition of sulfuric acid to produce hydrogen. The high temperatures that come from nuclear plants, especially the next generation nuclear plants (NGNPs) which will be high temperature gas cooled reactors (HTGRs), can be used to provide enough heat at a high enough temperature to decompose the sulfuric acid. The electricity that is produced from the HTGR can be used in electrolysis of the decomposition products plus other reactants to form hydrogen. Much research has been done on providing the required heat for decomposition and the physical constraints of how to transfer the heat from the NGNP to the sulfur process. Because of safety and contamination concerns, the hydrogen production plant must be significantly separated from the nuclear plant, either by physical means or by distance. An additional concern was possible contamination of the reactor coolant with chemicals present in the sulfur process. It was therefore planned that the heat from the NGNP be transferred to an intermediate fluid, which then transports the heat to the sulfur decomposition process, the schematic of which is shown in Figure 1. The heat exchangers for this process have been studied elsewhere [1], and will only be touched on here. The present focus is on the intermediate heat transfer fluid.

Because of both safety and economic reasons, it was decided to separate the sulfur decomposition system from the nuclear power plant in the hydrogen production process through distance rather than physical barrier separation. Therefore, the heat transfer fluid would have to travel significant distances through an intermediate coolant loop to connect the HTGR primary intermediate heat exchanger (IHX) to the sulfur process secondary IHX. Connecting both processes over long distances while efficiently transferring heat is not a trivial task. One of the main concerns is choosing an effective intermediate loop coolant. Originally, three materials were studied: liquid metals, molten salts, and helium. Because of the relatively low boiling points of most metals having low melting points, the first option was ruled out. Therefore, the main focus of recent research, and the focus of the present study, is to compare the properties of molten salts and helium as intermediate heat transfer fluids.

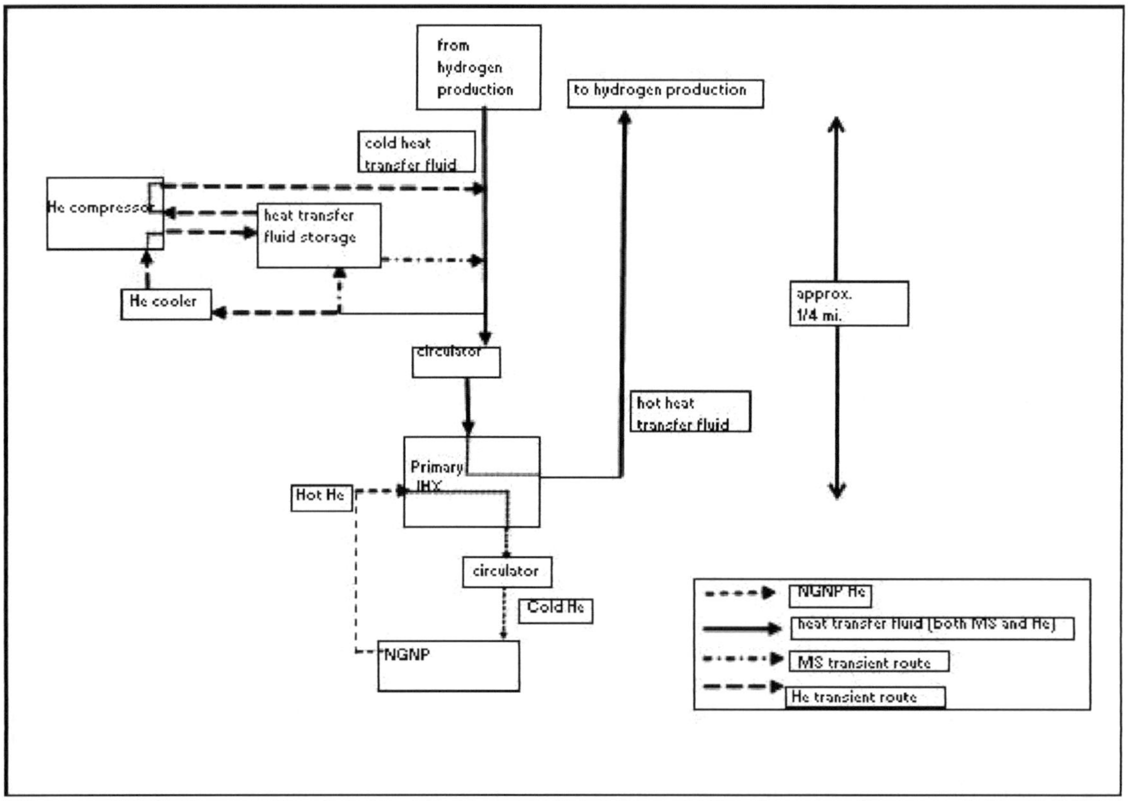

Figure 1. Schematic for the Intermediate Heat Exchange Loop using either Helium or Molten Salts.

2. Comparing Molten Salts

Molten salts were chosen as an intermediate loop fluid possibility for several reasons. Molten salts have good heat transfer properties, can be used at low pressures, have melting points in the optimal operating ranges, and have been studied for use in molten salt reactors. Although there are many molten salts available, it has been determined that fluoride salts are the most promising. [2] They have relatively low melting points, and are compatible with nickel or carbon based materials at high temperatures, although the compatibility at temperatures as high as the intermediate loop temperatures (~600-1000°C) is somewhat questionable. [3,4] A comparison of the properties of selected molten fluoride salts, as well as helium, is shown below. Although exact reasons for the recommendation are not clear, Oak Ridge National Laboratory feels that zirconium based salts are preferred over potassium based salts. [1] Because of this, and that less data seems to be available for potassium fluorides, it is recommended that they not be considered.

Table 1. Comparison of fluid properties at ~ 700°C (He at 10 MPa).

Coolant	Composition (mole %)	MP (°C)	Cp (J/g/°C)	density (g/cc)	ρ*Cp (J/cc/°C)	k (W/m/°C)	viscosity (Ns/m^2)	reference
Li2BeF4	-	459	2.34	1.94	4.5396	1	0.00563	3
LiF-BeF2	66-34	458	2.38	1.94	4.6172	1.1	0.0055	7
"	50-50	350	2.8	2.18	6.104		0.01076	4
LiF-NaF-BeF2	35-27-38	338	2.47	1.93	4.7671		0.0044	4
LiF-NaF-KF	46.5-11.5-42	454	1.88	2.02	3.7976	1	0.0029	4,7
LiF-NaF-ZrF4	42-29-29	460	1.47	2.71	3.9837	2.1	0.00444	3
NaF-BeF2	57-43	360	2.18	2.01	4.3818	1	0.00686	4,7
NaF-KF-ZrF4	10-48-42	385	1.09	2.83	3.0847	1	0.00159	2,7
NaF-ZrF4	58-42	500	1.17	3.14	3.6738	2.1	0.00166	3
He	-	-	5.2	0.00494	0.025688	0.29	0.00005434	1,5

As shown in Table 1, the pertinent properties of the different fluoride salts are very similar to one another. Therefore, the decision on which salt to choose depends on other factors not shown, such as stability of the salt. The stability of the salt will determine if it is going to be reactive with the container material and other chemicals, such as compounds present in the sulfur process. It was determined that lithium and sodium fluoride salts are most stable because of their low free energies.[2] Another factor is melting point. Depending on the operating parameters, there is a risk that the molten salt might freeze within the pipe. Different options to eliminate this problem have been explored, and will be discussed later. However, because of this risk, it is crucial to use salts with low melting points, especially if the distance between the primary and secondary IHXs is significant. In light of these facts, 50%LiF-50%BeF_2 and 35%LiF-27%NaF-38%BeF_2 seem to be the most promising molten salts, although more studies need to be performed.

It should be noted, however, that beryllium is known to be toxic. Beryllium fluoride is a known carcinogen, is very hygroscopic, and gives off beryllium fluoride fumes during melting. At normal loop operating temperatures, the vapor pressure from these fumes would create a concentration well above the maximum exposure limit. [16,17] Even if the salt was contained at all times, there would still be a possibility of leaking, and it would be unwise to allow for the health risk associated with the beryllium salt. Therefore, other salts that have less of a health hazard should be investigated for intermediate loop use. Most of the salts listed above with high heat capacities and low melting points all contain beryllium, and it will be a difficult task to find a salt with properties appropriate for this heat transfer application. The only exception might be NaF-KF-ZrF_4 with a melting point of 385°C, although this salt's heat capacity is lower than the heat capacities of beryllium compounds by around a factor of two. However, for the present, most calculations will use the salt 50%LiF-50%BeF_2 in order to present the maximum efficiency possible if using molten salts in the loop. One must keep in mind that even though the properties of the salts are different, they are along the same order of magnitude. The results from the following calculations will therefore not be drastically different if a different salt is used.

3. Comparing Helium with Molten Salts

In order to be an effective heat transfer fluid, the medium must be able to easily transfer heat with the primary HTGR coolant and the decomposition reactants, while not losing much heat in-between the two processes. As one can see from Table 1, the main two properties which govern the aforementioned constraints, the volumetric heat capacity and thermal conductivity, are orders of magnitude greater for the molten salts than for helium. This automatically makes them better candidates for the heat exchangers and intermediate loop. 50%LiF-50%BeF_2 in fact has the highest volumetric heat capacity of all of the fluids compared, at 6.1 J/cc/°C. However, before choosing an optimal heat transfer fluid, other comparisons must be made.

3.1 Transient operation

Because the sulfur process will have to increase or decrease its capacity, and even stop operation on a regular basis in order to meet the specific operating conditions of the reactor and the hydrogen process, the transient behavior of the intermediate loop must be considered. Changes in both temperature and pressure in the intermediate loop will ensue during startup or shutdown of the intermediate loop. This will change the amount of helium in the loop during transient conditions, although this will not affect the amount of coolant if the molten salt route is chosen. Helium storage tanks must therefore be able to accommodate these changes, as well as have enough helium to allow for gradual depressurization of the loop in case of a leak. A conservative estimate on the size of an auxiliary helium storage tank can be calculated. If the separation between the HTGR and sulfur process is ~ 400 m, the ID of the loop pipe is 0.4 m, and if the volumes of the heat exchangers and vaporizers are neglected, the loop volume would be 101 m^3, which would yield 17 kg of helium before startup (at standard atmospheric temperature and pressure). At the end of startup (before the sulfur cycle has begun), the helium temperature and pressure would be 950°C and 10 MPa, respectively, and would yield 395 kg of helium. However, at the end of pressurization (10 MPa) but before heat up the required helium would increase to 1,625 kg. This inventory, stored at ambient temperature under 34.5 MPa, would require a 29 m^3 (or approximately 7,700 gal) storage tank. If the diameter of the helium loop would be increased to 1 m (which is suggested in several studies [1,2] and will be discussed later), this volume jumps to 182 m^3 (48,100 gal). Keep in mind this is not including the helium in the heat exchangers or the helium required for an accidental leak in the loop.

A molten salt storage unit would only have to account for small changes in salt volume and would not have to account for leaks of the salt, and it would require a smaller volume than the helium storage unit because of the smaller pipe diameter in the loop (discussed later). If neglecting the heat exchanger volume and taking the diameter to be 0.0875 m, the volume of the loop (and consequently the volume of molten salt) would be 4.81 m^3, or 1,271 gal. Even if this diameter is increased to 0.23 m [1,2], the volume of the storage unit would have to accommodate 33.2 m^3 (8,781 gal.) of molten salt, which is over five times less than the required helium volume from sources 1 and 2.

The helium that is in storage can then be introduced into the intermediate loop without any physical modifications (except to depressurize the gas to the loop conditions- 10 MPa), and can continue to circulate until it is heated in the primary IHX to the optimal operating temperature, at which time the sulfur process can be started. Although it is unnecessary for the helium to be preheated during startup, a compressor is needed for the high operating and storage pressures. By comparison, salt stored at room temperature would be a solid, and would have to be heated in the storage unit until it melts. Furthermore, since there is heat loss in the loop, the fluid would have to be heated past the melting point before introduction to the loop to prevent freezing in the pipes. A heater for the molten salt storage unit is therefore needed to melt the salt during start up and to prevent the salt from freezing during short reactor outages. If the hydrogen process is down for a short time, a flow of molten salt can be kept moving in the IHX loop to maintain it at temperature.

A loss of power could cause freezing, but in a much more inconvenient way. If the circulator power is lost, molten salt will be stuck in the pipe and will eventually freeze. It would probably be difficult, or perhaps not possible at all, to melt all of the salt in the pipes, not to mention the heat exchangers. Additionally, if melting was possible, the thermal expansion of the salt could cause pipe rupture. Therefore, the two most reasonable remedies are to route alternate power to the circulator, or to place heat tracing around the pipe. The heat tracing would have to be designed to account for the massive thermal expansion of the pipe. Since it would be too expensive to run the tracing for an extended amount of time (e.g. until the circulator power is restored), a drainage system would have to be designed in the pipe to allow for removal of the salt into storage units which can more efficiently melt the salt upon freezing. If circulator power is lost, the heat tracing could be turned on to prevent freezing just long enough to allow for drainage into the storage tanks. Preliminary ideas of drainage systems are shown in Figure 2. The heat tracing would also be used to preheat the pipe to prevent initial salt freezing upon start up.

A main concern (which has been briefly mentioned) that will arise with transient operation will be the change in pipe temperature. Even if the material that is chosen for the loop pipe has a very low coefficient of thermal expansion, the massive temperature increase during startup and decrease during shutdown (possibilities of ΔT being ~900°C) will cause the pipe to significantly expand and contract. If a correct loop system is not built, the possibility of pipe cracking or permanent deformation is very high. Although the loop would be mechanically designed to account for thermal expansion, rapid increase in temperature by either molten salts or helium would lead to high stress in the pipe material. If the loop were heated at a slower rate, much stress and strain could be eliminated. One option that could do this is to run warm gas through the pipe and gradually increase the gas temperature. This option would be extremely easy if helium was used as the heat transfer fluid; the cold initial gas would simply be circulated in the loop with the nuclear reactor running but without the sulfur process running until it got up to temperature. If molten salt is chosen, a separate design would have to be created to input gas into the loop and remove it when the pipe got up to a reasonable temperature. This method would also help to prevent molten salt freezing upon start up. However, it could do nothing to prevent freezing if the circulator power was lost, so heat tracing would still have to be installed. Therefore, it should be decided if it is economically beneficial to install and run an auxiliary gas heating system rather than to run the heat tracing when initially warming the loop.

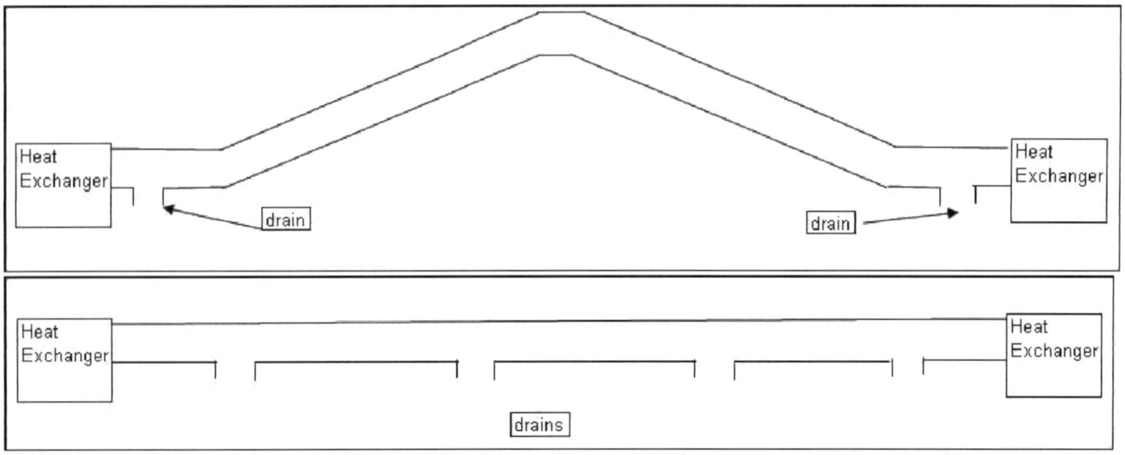

Figure 2. Possible drainage design systems for emergency salt evacuation.

3.2 Steady state operation

During steady state operation, the two most important factors to consider when choosing the optimal heat transfer fluid are heat loss and pumping power. To meet the energy requirements for the secondary IHX (decomposition reactor), certain flowrates for each fluid must be achieved. Calculations were completed and the results are shown in Figure 3. This is only a comparison, being that the ΔT of the fluid and the decomposition reactor heat duty were assumed (340°C and 32.2 MW). The mass flow rates of the coolants are relatively similar. However, as one would expect, the required volumetric flow rate of helium is orders of magnitude higher than that of the molten salt. Using this data, heat losses, pumping power requirements, and optimal pipe diameters can be calculated. These three parameters are based primarily on calculations concerning heat loss through the walls of the pipe. Because this is the cornerstone of the present study, a brief explanation on the method of calculations is presented.

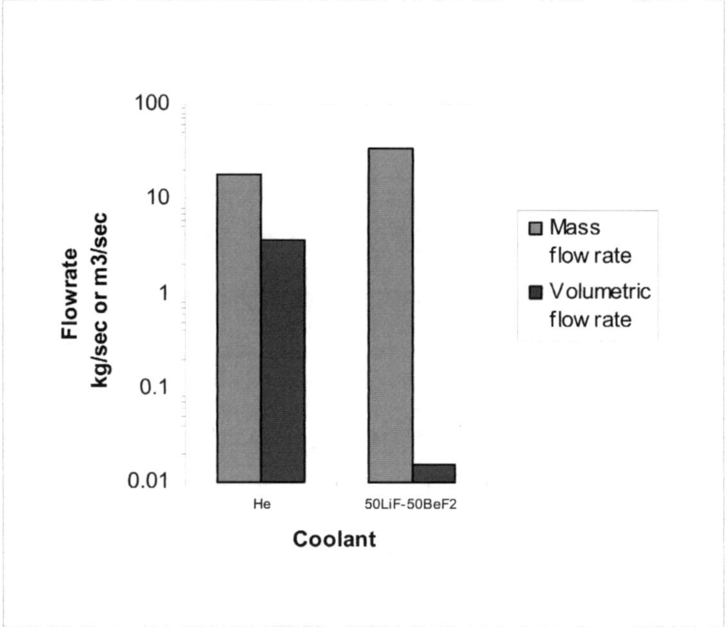

Figure 3. Flow Comparison to Meet Decomposition Heat Requirements (He at 10 MPa and 700°C).

Helium at 10 MPa and 50%LiF-50%BeF_2 were used as the heat transfer fluids. The determination of fluid flow rate was previously explained. Because it has been suggested that certain Haynes, Hastelloy, and Inconel alloys are possibilities for loop materials of construction [1], their thermal conductivities were averaged. This average, 28 W/(m/K), was used in this study. Because molten salts can be operated at low pressures, the pipe walls do not have to meet thickness requirements, unless corrosion and external and high temperature stresses are considered. Therefore, to be conservative, the thickness of the pipe walls for molten salt transport was 2.54 cm. The wall thickness required for 10 MPa helium can be calculated by

$$t = P*OD/(2*(S*E+P*Y)) + C \qquad (1)\ ([11])$$

where t is the minimum thickness of the pipe in inches, P is the internal pressure in psi, OD is the outer diameter in inches, S is the allowable stress in psi, E is the longitudinal weld joint factor, C is the sum of allowances for corrosion, erosion, and thread or groove depth in inches, and Y is a constant specific to the material. Again, the constant was chosen to be the most conservative. The insulation thermal conductivity and thickness were chosen to be 0.28 W/(m/K) (the reason of which will be explained later) and 5.08 cm, respectively. This 5.08 cm is an arbitrary number, and in several calculations different thicknesses were used for comparison purposes. Using these parameters and fluid properties obtained from numerous sources [1, 3-6, 12], convective, radiation, and conductive heat transfer rates were calculated. Because the temperature of the fluid is constantly changing throughout the ½ mile length of pipe, the heat transfer rate is constantly changing which makes it extremely difficult to find a solution analytically. What was done to get around this problem was to separate the pipe into 1 meter length sections. If the heat flux through this small section is assumed to be constant, a solution can be found. The lost heat will drop the fluid temperature, which will lead to a slightly smaller heat flux in the next section. In this way the total heat loss and temperature drop throughout the whole loop was calculated.

Assuming that the pressure drop for both heat exchangers was 112 kPa and the pipe equivalent roughness was 0.15 mm, total energy loss (heat loss through the pipe walls + thermal pump power required) was minimized by varying the pipe diameter. This was done for a number of inlet temperatures (being the temperature of the heat transfer fluid right out of the primary IHX), and a set of optimal operating parameters was obtained. The results are shown in Figures 4 and 5. The optimum diameters for each fluid do not change when the inlet temperature is changed. As one can see, pipe inner diameters which yield the lowest losses for helium and molten salt are 0.4 m and 0.0875 m, respectively. This agrees with earlier predictions that helium loop diameters would be 5 times that of molten salts [3, 7], although diameters are lower than previous studies [1,2] by a factor of 3.

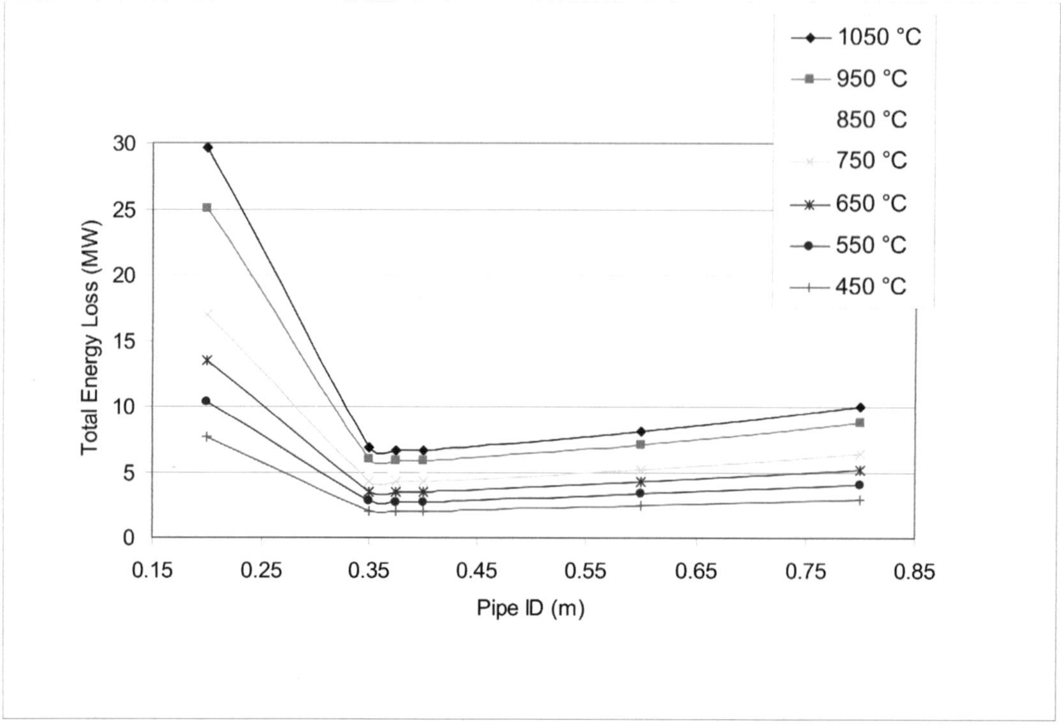

Figure 4. Optimal Diameter for Helium Transport at Varying Intermediate Loop Inlet Temperatures.

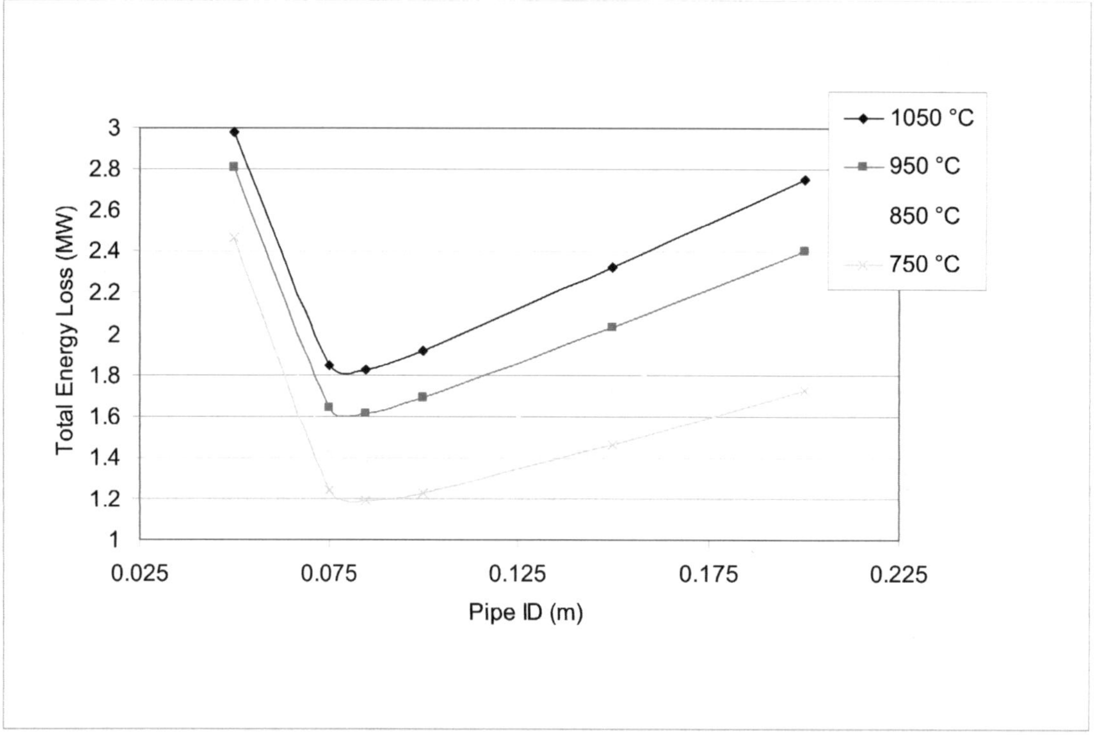

Figure 5. Optimal Diameter for MS Transport at Varying Intermediate Loop Inlet Temperatures.

The chosen diameters were used to calculate total energy losses, which are displayed in Figures 6 and 7. Although no temperature constraints were held for helium, one must consider freezing when molten salt loops are modeled. 750°C seems to be the lower limit for loop inlet temperatures for molten salts. Lower temperatures would result in molten salt solidification before the fluid would reach the primary IHX. Keep in mind that the study used 50LiF-BeF_2 as the salt, and that most salts have higher melting points, which would limit salt temperature even more.

Figure 6 shows relative total energy loss for both heat transfer fluids with a 32.2 MW decomposition reactor heat duty. This, more than anything, shows the molten salt's superiority over helium in heat transfer, and why helium would be extremely inefficient in the loop. The graph suggests that at the normal operating temperature of 950°C, 15% of the total heat transferred in the helium loop comes from losses, however this number is decreased by a factor of three if molten salt is used. At 950°C, doubling the insulation thickness (from 5.08 cm to 10.16 cm) seems to decrease the total energy loss to ~11.2% for helium and ~3.5% for molten salts. A break down of the heat transferred is shown in Figure 7, and from this the heat that the primary IHX must transfer to obtain the original fluid temperature was calculated, and is shown in Figure 8.

Although the heat loss should be minimized, other factors must be considered. Sulfuric acid only can decompose above a certain temperature. Some studies suggest that this minimal temperature is 850°C [2]. Figure 9 shows the temperature of the fluid reaching the decomposition reactor. The minimum loop inlet temperatures to reach 850°C are ~852°C and 872°C for the molten salt and helium, respectively. Again, this and mostly all other calculations were based on the original design parameters; the increase in insulation thickness to 4 inches was for comparison purposes only.

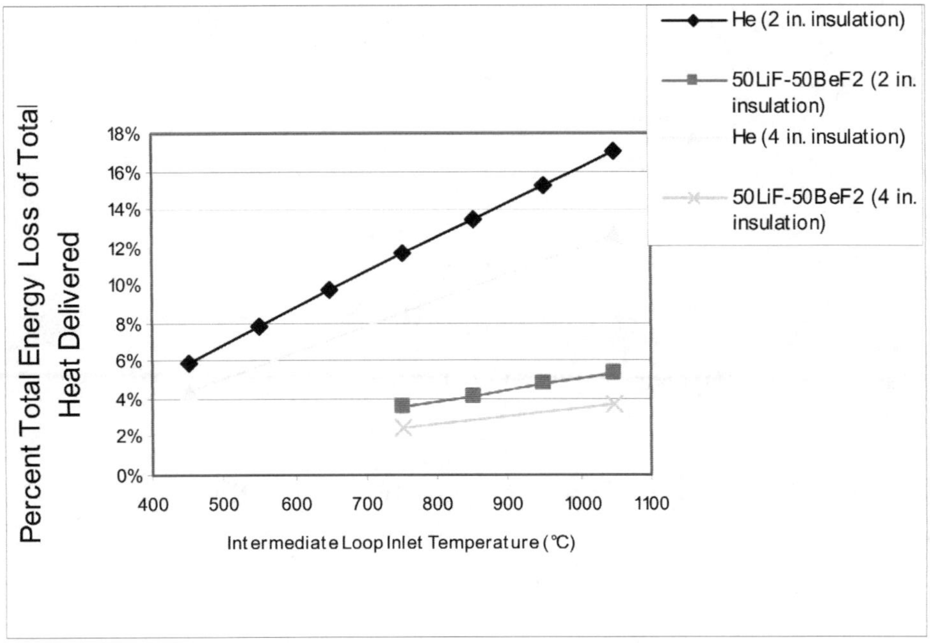

Figure 6. Comparison of Total Energy Loss for Different Fluids in the Intermediate Heat Transfer Loop at the Optimum Diameter.

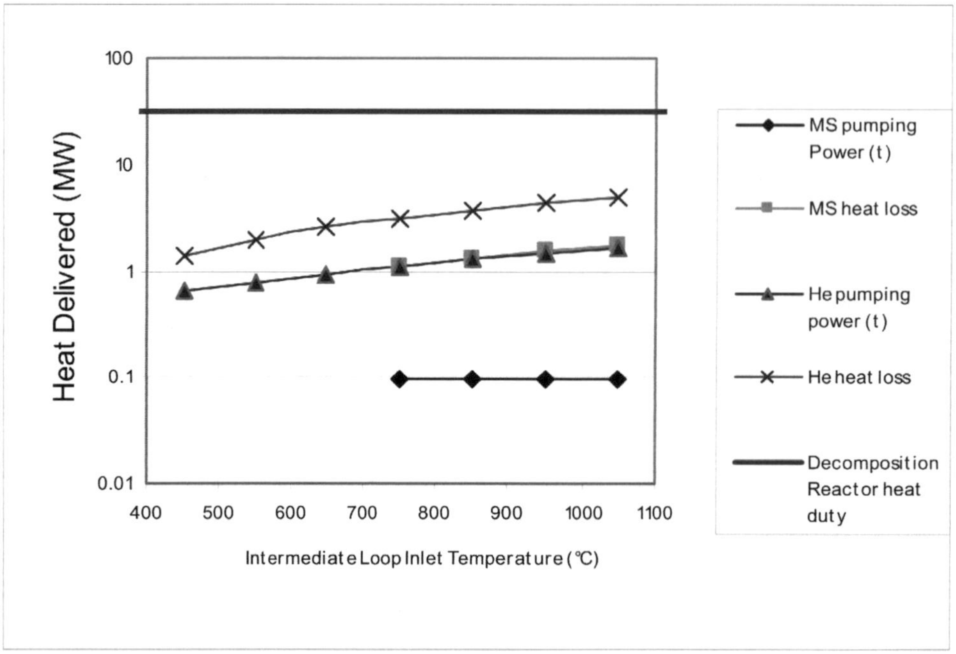

Figure 7. Heat Transferred vs. Temperature at Optimum Diameter.

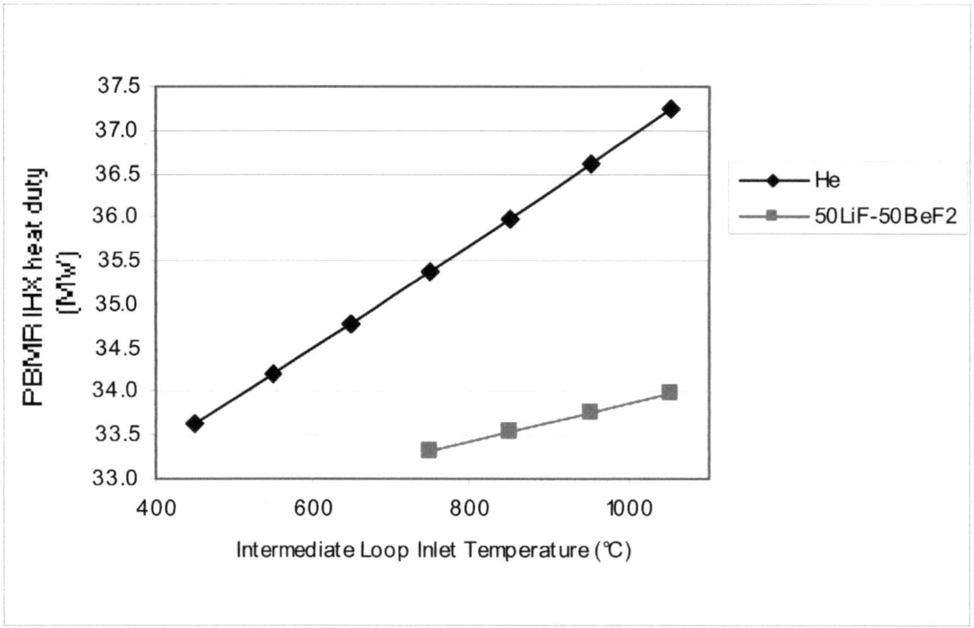

Figure 8. Required PIHX Heat Duty at Optimum Diameter.

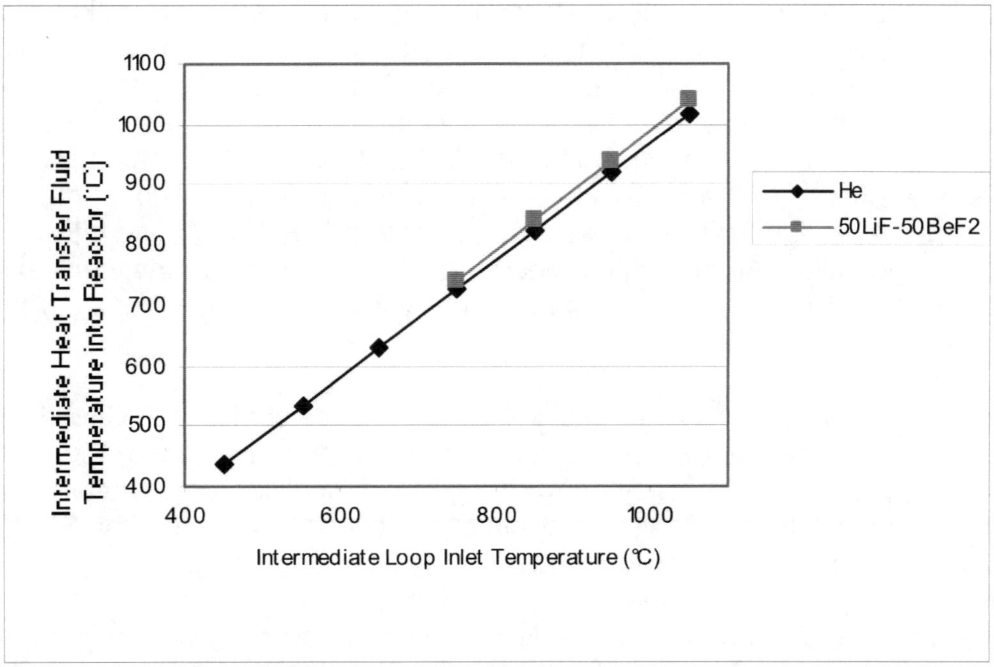

Figure 9. Variation of Decomposition Reactor Inlet Temperature with Loop Inlet Temperature at Optimum Diameter.

3.3 Materials

The conditions that arise with running the intermediate loop make choosing the materials of construction a difficult task. For helium, a concern is the strength of the material not only at elevated temperatures but elevated pressures as well. For the molten salts, a possible issue will be compatibility with the heat exchanger and pipe material. Because the loop is run at such a high temperature, creep rate and thermal expansion are significant, as well as corrosion and oxidation, for any heat transfer fluid chosen. Other issues arise, and will be addressed here.

3.3.1 Compatibility

Because molten salts have been looked at for use in advanced reactors, containment materials for the salts have been studied. If the fluoride potential is kept low, a reducing environment is maintained, and most importantly, the correct pipe material is chosen, corrosion from fluoride salts should not be a problem. [3] It is known that chromium is one of the more soluble elements in most alloys when fluoride salts are used. [3] Most acceptable materials therefore have a low concentration of chromium. A more thorough study investigated corrosion rates of numerous fluoride salts with several different metals. [13] They found that the tendency for the investigated metals to corrode in fluoride salts is (in increasing order) as follows:

$$Ni<Co<Fe<Cr<Al$$

This agrees with other studies [3,4,14] which recommend the use of high nickel alloys (with minimal chromium).This study also found that LiF is the least corrosive fluoride salt, which also agrees with previous studies. This study performed equilibrium calculations for 1 mole of metal in molten salts at 1100 K. It was shown that the concentrations of NiF_2, CoF_2, FeF_2, and CrF_2/CrF_3 were much less than 1 mol ppm in the LiF melts ($10^{-3} - 10^{-0.5}$ mol ppm), and the concentrations of the first three were also low in NaF and KF (0.032-1 mol ppm). Chromium corrosion, however, was significant in these two melts, and CrF_2/CrF_3 concentration was relatively high (16 mol ppm). Aluminum fluoride compound concentrations were extremely high in all three melts ($10^3 - 10^5$ mol ppm). It would be beneficial to perform these studies using the salts listed in Table 1.

Haynes 230 and Inconel 617 are being studied as possible Printed Circuit Heat Exchanger (PCHE, which will be discussed momentarily) materials [1], and therefore have possibilities for the loop. Most Hastelloy alloys have either Ni or Mo as a major constituent, and therefore have been considered. However, some of the alloys (e.g. Hastelloy B) have poor oxidation resistance and age hardening characteristics at high temperatures [4]. One solution to this problem is to incorporate Inconel with the Hastelloy-B. Inconel improves corrosion and high temperature characteristics. Inconel's relatively high chromium content is the reason why it cannot be solely used. One hybrid that has been studied which contains these two alloys is INOR-8, which has good compatibility with fluoride salts and good oxidation resistance. Corrosion studies that were done with the alloy at 700°C show "excellent corrosion resistance to molten fluoride salts." [4] The study reports that at these conditions the oxidation rate was <0.0508 mm in 100,000 hrs. However, more studies are suggested at higher temperatures before the alloy is seriously considered. Normal melting procedures can be used to prepare INOR-8. The material can also be hot forged using a number of techniques, including extrusion and hammer and press techniques. [4]

One of few major alternatives to nickel based alloys is carbon-carbon composites. Carbon-carbon composite materials were designed to have high strength at temperatures greater than 1650°C, a low coefficient of thermal expansion, and oxidation resistance. As opposed to ceramics and superalloys where tensile strength goes down when temperature goes up, the tensile strength of carbon-carbon composite materials stays constant, and is generally greater than ceramics and superalloys. Properties of typical carbon-carbon materials are shown in Table 2. [15] A possibility is to coat the inside of the heat transfer loop and heat exchangers with carbon. It is thought that molten salts will not corrode chemically vapor infiltrated carbon coats, the coating will maintain full mechanical strength, and will not be subject to creep. [1]

Table 2. Typical carbon-carbon composite properties.

Tensile Strength	>276 MPa
Melting Point	>4100°C
Modulus of Elasticity	>69 GPa
Thermal Conductivity	11.5 W/(m/K)
Linear Thermal Expansion Coefficient	1.1E-6/°C
Max. T in Oxidizing Environments	1600°C

Although compatibility is not any issue when dealing with helium, material strength is more of a concern. Carbon-carbon composites can still be an option because of their excellent high temperature properties, although their resistance to helium permeation is still questionable. If either carbon-carbon materials or alloys are chosen for helium transport, they must have low thermal expansion coefficients, low creep, and high strength at helium pressures and temperatures.

It should be noted that one study suggests that air corrosion is usually more of an issue than molten salt corrosion, especially at temperatures above 800°C. [7] This would mean that as far as corrosion is concerned molten salts would be no worse than helium in the loop.

The same consideration must be given when choosing storage unit materials as when choosing piping materials. The only difference is that the helium will be at higher pressures (~34.5 MPa). This will require different materials to handle the additional stress. Also, because of this high pressure storage, a helium cooler will have to be coupled to the storage tank. Should the cooler fail, the pressure inside the auxiliary storage unit would increase dramatically, and the unit must be stable enough during a worst case scenario. Many high strength materials that would be used for the storage tank are not structurally sound at the storage pressure and temperatures of the intermediate loop. Material of construction for the storage tank should therefore be carefully chosen.

The molten salt storage container would be no different than the piping, except that a preheater must be designed for the melting of the salt. The material requirements, however, can be less stringent because the molten salt will generally be at a lower temperature than the loop, so the salt will be less likely to react with container materials.

Along with the fluid compatibility with materials of construction, it must be compatible with process chemicals. One of the biggest concerns of using molten salts for a heat transfer fluid is the possibility of a leak from the loop into either of the heat exchangers. On the primary end, a leak would mean molten salt into the primary helium coolant and possible transfer into the advanced reactor. On the other end of the loop, a leak would mean possible reactions with sulfuric acid decomposition reactants and products. The former situation would lead to shutdown of the reactor. The latter situation can have numerous consequences. Hot fluoride salts in contact with water have the possibility to cause steam explosions and to produce HF. To test this, 230 kgs of molten salt was heated to a temperature of 815°C and injected into water for a period of 45-50 seconds. [7] No steam explosion was reported, although it was not reported whether or not HF formed or even if the salt was fluoride based. Thermodynamic calculations were done for the system of $NaF\text{-}KF\text{-}ZrF_4$ and sulfuric acid decomposition products (H_2O, O_2, SO_2, and SO_3). [2] The reaction is endothermic, and results in a 15% increase in gas volume. Fluoride salts reacting with these decomposition products, as well as sulfuric acid, will yield HF, and more thermodynamic and kinetic studies should be attempted to find the amount and rate of HF generation at different conditions. If the salt contains some water in the fabrication stage (>10 mol ppm H_2O can cause HF formation [13]), it can normally be removed by heating *in vacuo* if the particular salt is not easily hydrated. [16]

Although no reaction could happen between helium and the process chemicals, failure anywhere in the helium loop would cause a change in pressure in the plant. A pressure increase in any part of the plant would lead to the possibility of massive chemical release if the pressurized helium would cause an explosion and transfer chemicals through the air. This possibility is further worsened by the fact that the helium in the proposed loop would be pressurized to 10 MPa. Storage facilities must be able to accommodate for enough helium to enter the loop to cause gradual depressurization if there would be a loop leak or failure. The chemical industry in Germany actually discourages the use of helium for heat transfer loops when toxic chemicals are involved for this reason. [7] This is not a concern with molten salts because the fluid would not be pressurized and a leak in the system would not cause an extreme pressure difference in the plant.

3.3.2 Insulation and Heat Exchangers

It has been mentioned before that even if several centimeters of insulation (this study used ~5 cm) are placed on the loop heat loss is still very significant. It is therefore imperative that the correct insulation material be chosen. The most critical factor is of course thermal conductivity. Because of the huge temperature gradient between the fluid and air, especially right after the primary IHX, conductivities must be extremely low. Although there are many insulating materials available, few are able to insulate or even exist at the temperatures present during normal loop operation.

Table 3. Comparison of the properties of different insulating materials. [9,10]

Material	density (kg/m3)	Max operating temperature (°C)	k (W/m/K) (@1100°C)
Micro-quartz fiber, blanket	48	1649	0.25
Zirconia grain	1810	1649	0.38
Gemcowool™	192-256	1150	0.23 (@ 980°C)
Gemcolite ™	208-288	1260	0.28

Calculations have shown that even with relatively low conductivity (~28 W/(m/K)) pipe material, the temperature of the pipe/insulation interface can be only 10s of degrees lower than the transfer fluid temperature. This means that depending on the initial temperature, the insulation could get very hot. Although ceramics were considered because of their high operating temperature and low conductivity, the latter property was not low enough to limit heat loss to an appreciable extent. Table 3 lists possibilities of high temperature insulating materials. Gemcolite^{TM}, which is made by RSI (Refractory Specialties Inc.), was chosen as the insulation in the current study's calculations (using a constant thermal conductivity of 0.28 W/(m/K)) for the fact that the RSI products were made specifically for high temperature conditions. [9] Before any insulating material could be chosen, however, it would have to be determined if the material could be shaped around the loop pipe and if any reaction would ensue with the insulator and the pipe material.

Research of the optimal primary IHX has be done elsewhere [1], so only a brief summary will be presented here. The heat exchanger used to transport heat from the primary HTGR coolant to the intermediate fluid will subjected to the same issues as the loop itself. However, there are two properties that limit the use of the heat exchanger: heat transfer and pressure drop. At the present time, it is thought that the Printed Circuit Heat Exchanger (PCHE) will be the best primary IHX. The PCHE, which is now manufactured by Heatric, implements semi-circular flow channels etched onto metal plates. These plates are stacked together and heat transfer ensues between the channels. The PCHEs are not only capable of handling high pressures; they are made to handle high pressure drops as well. Also, as previously mentioned, if helium is used as the transfer fluid and there is a leak in the system, the loop must be gradually depressurized. It is thought that the PCHE could be one of the few heat exchangers that could handle this kind of depressurization. [1,8]

The efficiencies of the decomposition reactor and the primary IHX are both affected by the type of heat transfer fluid chosen. Since it has been suggested that overall heat transfer coefficients for He-He and He-MS heat exchangers are along the same order of magnitude [1], other comparisons should be made. Although it has been mentioned that PCHEs have been known to withstand pressure drops of the magnitude that could be seen if helium is used, minimal stress is of course always preferred. If molten salt (which can be run at any pressure) is used, the IHX would be subjected to much less stress. A lack of pressure constraints also gives more design freedom for the heat exchangers, which can ultimately lead to better efficiency.

Studies have been done showing the relationship between the logarithmic-mean temperature difference (LMTD) of the heat exchanger fluids and pumping power and volume for the intermediate heat exchanger. [3,7] If the LMTD is the same for both helium and molten salt, the power required to pump the fluid through the heat exchanger, as well as the required volume of the heat exchanger, is much higher for the gas then the liquid. As the LMTD goes down, the difference in HX volume for the two fluids increases drastically.

4. Conclusions and Recommendations

In light of the present study, it is believed that molten salts will be the optimal heat transfer fluid for use in the intermediate heat exchange loop. Several molten salts were compared with helium. Fluoride salts were chosen because of their high temperature compatibility with materials as well as their relatively low melting points. Because the transport properties of most of the fluoride salts considered were very similar, the salt 50%LiF-50%BeF_2 was chosen due to its melting point (~350°C). Also, its volumetric heat capacity was higher than the rest of the molten salts, which is favorable in the heat transfer loop.

Transient and steady state conditions and requirements were then compared between 50%LiF-50%BeF_2 and helium. The majority of this information was based on calculations concerning fluid flow and total energy loss through the loop pipe. These calculations showed that total energy loss for helium is over 3 ½ times greater than that for the molten salt, and can reach up to 18% of the total heat transferred depending on the starting temperature of the loop, which is mainly due to the high helium pumping power requirements. This is the main characteristic that makes molten salts more appealing than helium for heat transfer.

Materials of construction were also compared between the helium system and molten salt system. The materials that are recommended for molten salt transport are high nickel alloys with minimal chromium or carbon-carbon composites. Below is a summary of the pros and cons of each heat transfer fluid.

Table 4. Final comparison of helium and molten salt heat transfer fluids.

Fluid	Pros	Cons
Helium	No preheating required, Freezing not an issue, Reactions between it and process materials/chemicals not an issue	**High pressure required which could lead to high chemical release, Large flow rate requirements and large compressor, pump, and heat losses,** Large storage volume needed to account for leaks and temperature changes, May cause HX failure if ΔP is large
MS-50%LiF-50%BeF2	**Smaller flow rate and diameter requirements and smaller heat and pump losses,** Compatible with several containment materials, Efficient in the heat exchangers	Toxic, Freezing possible, Storage tank heater required, Possibly reactive with process chemicals and materials

Because of the much higher energy losses associated with helium use, it is recommended to use molten salts for the heat transfer fluid. Although 50%LiF-50%BeF_2 is recommended because of its low melting point and high volumetric heat capacity, it has not been a very well studied salt, so more studies on the compound should ensue. The salt will toxic because of its beryllium content, so other salts (possibly zirconium based with the same characteristics as mentioned above) should also be investigated. Studies of the salts should include their stability, purification, fabrication, handling, and compatibility with materials of construction. The most important of these studies should be determining the reactivity of it and the chemicals in the nuclear and hydrogen processes.

References:

[1] L. Paoletti and E. J. Lahoda. "Intermediate Heat Exchanger Coupled to a Generation IV Reactor." Westinghouse Science and Technology Center internal document, Dec. 2004.
[2] C. Forsberg and P. Pickard. "Alternative Heat Transport Systems for High-Temperature Heat." ORNL, DOE Independent Technical Review Group. February, 2004.
[3] P. Peterson, H. Zhao, and G. Fukuda. "Comparison of Molten Salt and High Pressure Helium for the NGNP Intermediate Heat Transfer Fluid." University of California, Berkeley, Dec 2003.
[4] J. A. Lane, H. G. MacPherson, and F. Maslan, Eds. Fluid Fuel Reactors. Addison-Wesley Pub. Co., 1958.
[5] I. Barin. Thermochemical Data of Pure Substances. VCH Verlagsgesellschaft mbH, D-6940 Weinheim (Federal Republic of Germany), 1989
[6] W. Lick and H. Emmons. Thermodynamic (and Transport) Properties of Helium to 50,000 K. Harvard University Press, Cambridge, MA, 1962.
[7] C. W. Forsberg. Reactors with Molten Salts: Options and Missions. ORNL. The 2004 Frederic Joliot and Otto Hahn Summer School, 2004.
[8] E. J. Lahoda, R. Matzie, and A. Botma. "Interfacing the Westinghouse Sulfur Cycle with the PBMR for the Production of Hydrogen." Westinghouse Science and Technology Center internal document, February 2004.
[9] *Ceramic Fibre Shaped Insulation; Refractory Specialties Inc.* Refractory Specialties, Inc. Accessed January 2005. http://www.rsifibre.com.

[10] T. Baumeister E. A. Avallone, T. Baumeister III, Eds. Marks' Standard Handbook for Mechanical Engineers, 8th Ed. McGraw Hill Book Company, New York, NY, 1978.
[11] R. H. Perry and C. H. Chilton, Eds. Chemical Engineers' Handbook, 5th Ed. McGraw Hill Book Company, 1973.
[12] J. R. Welty, C. E. Wicks, R. E. Wilson, and G. Rorrer. Fundamentals of Momentum, Heat, and Mass Transfer, 4th ed.. John Wiley and Sons, Inc., 2001.
[13] A. K. Misra and J. D. Whittenberger. "Fluoride Salts and Container Materials for Thermal Energy Storage Applications in the Temperature Range 973-1400 K." NASA Technical Memorandum 89913 AIAA-87-92226. 22nd Intersociety Energy Conversion Engineering Conference, Philadelphia, Pennsylvania, 1987.
[14] Proposal-Development of Advance High Temperature Heat Exchangers. To the U.S. DOE from UNLV Research Foundation. 2003.
[15] S. T. Peters. Handbook of Composites, 2nd Ed. Kluwer Academic Publishers, 1998.
[16] G. J. Janz. Molten Salts Handbook. Academic Press, New York, NY, 1967.
[17] Brush Wellman Engineered Materials. Material Safety Data Sheet: No. M12, Beryllium Fluoride, January 13, 2003.

73c Model-based Performance Comparison of Thermochemical Nuclear Hydrogen Processes

Maximilian B Gorensek

Evaluations of the thermal efficiency of several different flowsheets for the production of hydrogen from nuclear energy being developed for the Nuclear Hydrogen Initiative (NHI) have been undertaken at the Savannah River National Laboratory (SRNL), using steady-state process models. The ultimate goal of this work is to provide an objective determination of the true thermal efficiencies and capital costs of the leading NHI alternatives. This information will be crucial for making the best choice of which NHI development path(s) to follow. The approach has been to build an "apples-to-apples" comparison, using a consistent basis, e.g. identical nuclear heat source, H2 and O2 as sole products, etc. Using all of the available information, complete flowsheets are designed, including the nuclear heat source and accounting for all utilities. These flowsheets are then used to create steady-state Aspen Plus models from which material and energy balances can be drawn. Energy utilization is subsequently optimized and the net thermal efficiency finally determined. Cost estimates could be added as a further refinement of these analyses.

To date, only the Sulfur-Iodine and the Hybrid Sulfur processes have been analyzed. The presentation will provide an up-to-date assessment of the status of this ongoing effort.

73d H2 Production Using High Temperature Steam Electrolysis Supported by Advanced Gas Reactors with Supercritical CO2 Cycles

Bilge Yildiz, Katherine J Hohnholt, Mujid S Kazimi

A design evaluation has been performed for an integrated system of high temperature steam electrolysis (HTSE) supported by a supercritical CO2 (SCO2) recompression Brayton cycle that is directly coupled to an advanced gas cooled reactor (AGR). The analysis includes the study of major components performance parametrically and the component integration systematically for different plant configurations. The configuration (HTSE-SCO2-AGR) with thermal recuperation from the gas product streams and an intermediate heat exchanger (IHX) between the turbine exit and the feed water stream is expected to be superior with respect to thermal efficiency, operational flexibility and cost.

HTSE for H2 production provides two major advantages: 1- It can promise higher efficiency at intermediate temperatures (with the appropriate materials and power conversion system for intermediate temperature operation of the solid oxide electrolysis cell) compared to the leading high temperature thermo-chemical water splitting cycles, 2- It has already been demonstrated experimentally by several groups, based on the materials used for solid oxide fuel cells. Thus, there is margin for improving its performance further with processing of new materials that can minimize the energy losses due to the electrocatalytic mechanisms specific to electrolysis on oxide electrodes. HTSE energy efficiency depends strongly on the electrical efficiency. The supercritical CO2 recompression Brayton cycle is a strong candidate to enhance the HTSE efficiency due to its major advantage of having high electrical efficiency (comparable with that of He cycle) at significantly lower temperature (550oC vs 850oC). In addition, it is expected to reduce the capital cost of the plant due to its simplicity and compactness. The advanced gas cooled reactor, AGR, which has extensive operational experience, is chosen as the nuclear reactor that can support the SCO2 cycle and provide heat for the HTSE in this design. The SCO2 power conversion system is proposed to couple to the AGR with a direct cycle. This requires a design update on the AGR to withstand high pressures at operation near 700oC. Nevertheless, it can reduce the overall cost of the plant and improve the system efficiency. An intermediate heat exchanger that provides heat from the CO2 at the exit of the turbine to boil the feed water stream for electrolysis is proposed for operational flexibility of adjusting the hydrogen and electricity output rate. In addition, the IHX at this configuration is subject to less harsh temperature environment compared to one that can be placed at the exit of the reactor.

The HTSE average process temperature is chosen as 900oC and the hydrogen pipeline delivery pressure is assumed to be 7MPa for the evaluation of each design configuration in this work. The reactor exit temperature and the SCO2 cycle turbine inlet temperature are chosen the same as those for the SCO2 recompression cycle design options for consistency and compatibility. For the base case design, the reactor exit temperature is selected as 550oC, the HTSE operates at 1Atm with 80%-90% voltage efficiency, the SCO2 cycle is based on the conservative design and the produced hydrogen is compressed for pipeline delivery. For these operating conditions, the system can achieve 43.5%-40.4% (LHV) overall hydrogen production energy efficiency, respectively. When the operating pressure of the HTSE is increased to 3MPa, the system can achieve 42.4%-48.2% (LHV) of hydrogen production energy efficiency, where the range depends on the SCO2 cycle operating conditions and the HTSE materials performance. With improved strength of the recuperators and HTSE cells, 7MPa can be expected as feasible product exit pressure from the HTSE system in the future. The design with HTSE operating at 7MPa can achieve energy efficiency of 41.6%-47.4% (LHV) and eliminates the hydrogen compressor before the delivery of the product into the pipeline. The product gas streams remain at relatively high temperatures even after the thermal recuperation in this design. Further thermal recuperation for bottoming a steam cycle, for preheating the water or for providing heat for the SCO2 cycle can be energetically possible. However, none of these alternatives are found viable economically. Process heating by the final product can be possible depending on the proximity to the necessary industry, and is not pursued in this study.

As with all new technology concepts, there are major R&D requirements with the recuperative HTSE-SCO2-AGR for its commercial realization. Some of the major needs for HTSE-SCO2-AGR have been identified as: materials processing for the durability and efficiency of the HTSE system at intermediate temperatures, the design update of the AGR with advanced materials to resist high pressure CO2 coolant, thermalhydraulics of CO2 at supercritical pressures, and detailed component design for system integration.

The energy efficiency of the recuperative HTSE-SCO2-AGR as analyzed in this work is more promising over the other leading nuclear hydrogen production technologies. Challenges for this system exist, but are less difficult than those for the very high temperature thermochemical hydrogen production processes proposed to be supported by future VHTRs that have more uncertainties in their development. In the near term, the aim must be to expand the research on this promising technology as summarized above, in order to have a system available for large scale hydrogen production with confidence and economy.

Hydrogen Markets: Implications for Hydrogen Production Technologies

Charles W. Forsberg
Oak Ridge National Laboratory
Oak Ridge, Tennessee 37831
Tel. (865) 574-6783; E-mail: forsbergcw@ornl.gov

Abstract

Different existing and future markets for hydrogen have varying characteristics that impact the competitive advantages or disadvantages of hydrogen generated by alternative technologies. Long-term stable markets favor higher-capital-cost, lower-operating-cost technologies. Dispersed small markets favor modular technologies, such as electrolysis, while large markets may favor centralized nuclear-hydrogen production options using thermochemical cycles that convert heat plus water into hydrogen and oxygen. Potential restrictions on greenhouse gas emissions may result in extremely large demands for hydrogen for liquid-fuel production and other uses, if the hydrogen production technology does not emit greenhouse gases to the atmosphere. Some hydrogen production technologies also create by-products such as oxygen. When a demand for such by-products exists, these by-products may also define the preferred hydrogen production technologies. Potential hydrogen markets, the characteristics of these markets, and some of the implications are described.

1. Introduction

The development of hydrogen production technologies requires identification of potential markets and the constraints associated with those markets. The markets partly define the requirements for the production technology as well as potential partners to commercialize new hydrogen production technologies. For example, if a potential user of hydrogen also required large quantities of oxygen, that user (if it were a partner) would only be interested in technologies that co-produce hydrogen and oxygen. Markets are not always easy to identify—particularly when the technology might not be fully deployed for several decades. This paper is an effort to identify potential major markets and some of the constraints associated with those markets. This provides a basis to allow assessment of the requirements for hydrogen production technologies.

The major future markets for hydrogen depend primarily upon four factors: (1) the future cost of hydrogen, (2) the rate of advances of various technologies that use hydrogen, (3) potential long-term restrictions on greenhouse gases, and (4) the cost of competing energy systems. For non-carbon-dioxide-emitting hydrogen production technologies (nuclear, renewables, and fossil fuels with carbon dioxide sequestration), potential restrictions on carbon dioxide emissions to the atmosphere are a key factor in the potential size of future markets. With the Kyoto Accord now in effect in most of the world (except the United States and Australia), there are growing incentives to reduce greenhouse gas releases. Such incentives already exist in some countries. For example, Norway taxes carbon dioxide emissions and thus creates a strong incentive to minimize these releases (Herzog and Golumb 2004). In any scenario where there are restrictions on carbon dioxide releases to the atmosphere and when non-greenhouse hydrogen production techniques at reasonable costs are available, extraordinary demands for hydrogen will likely exist.

2. Markets

Eleven potential major markets for hydrogen are considered, several of which represent alternative markets. Each market is characterized by several parameters that impact the choices of preferred production technologies as illustrated in **Table 1** and defined below.

- *Oxygen use*. Some markets require both oxygen and hydrogen. Such markets favor hydrogen production technologies that co-produce oxygen.

- *Heat use*. Some process applications require hydrogen and heat. This encourages co-location of the hydrogen production plant with the application.

- *Existing market*. Hydrogen is used in many markets. For existing applications, user technology is not the market constraint. If the cost of hydrogen is reduced, market use will increase.

- *Site specific*. Some hydrogen markets are unmovable. For example, hydrogen for aircraft must be delivered to airports. In contrast, refineries that require large quantities of hydrogen are built where there are low-cost sources of hydrogen. Today, most hydrogen is made from natural gas; thus, refineries that require large quantities of hydrogen are located near low-cost sources of natural gas. For site-specific hydrogen applications, alternative methods of hydrogen production become competitive when their cost is lower than that from natural gas at the specific location.

- *User size*. If a single large hydrogen consumer uses all the hydrogen from a production facility, development of such a facility is simple. However, if many customers are needed to consume the hydrogen from a production facility, major infrastructure elements (pipelines, etc.) are required to deliver the hydrogen as well as the added difficulty of matching production to demand. Markets with many small users are more difficult to commercialize.

- *Steady-state*. Some applications are constant users of hydrogen, while others require variable delivery.

- *Storage and distribution*. Different applications have very different needs for a supporting hydrogen storage and distribution system.

3. Descriptions of Markets

3.1 Liquid Fuels

About 40% of the U.S. energy demand is met by oil that is converted primarily to liquid fuels. Today's transportation system depends upon liquid fuels (gasoline, diesel, and jet) because of their high energy density by weight and volume and their ease of use. Liquid fuels can be made from hydrogen and any source of carbon (crude oil, heavy crude oil, tar sands, coal, biomass, and atmospheric carbon dioxide). Liquid fuels are typically hydrocarbons that have ratios of hydrogen to carbon of 1.5 to 2. Gasoline, diesel, and jet fuels are characterized by their performance in engines. No fixed hydrogen-to-carbon ratio is associated with each fuel.

Table 1: Markets for Hydrogen

Market	Product	O_2 Use	Heat Use	Exist. Market	Site Specific	User Size	Steady State	Store & Distribute
Refineries	Liquid Fuel	No	?[b]	Yes	No	Large	Yes	No
Heavy Oil & Tar Sands	Liquid Fuel	?[b]	Yes	Yes	Yes	Large	Yes	No
Coal Liquefaction	Liquid Fuel	Yes	Yes	Yes	Yes	Large	Yes	No
Atmospheric Liquid Fuels	Liquid Fuel	No	?[b]	No	No	Large	Yes	No
Biomass Fuel	Liquid Fuel	No	?[b]	No	Yes	Medium	?	Yes
Carbon Dioxide Recycle Fuel	Liquid Fuel/CaO	No	No	No	Yes	Medium	Yes	Limited Distribute
Chemical	Chemicals	No	?[b]	Yes	No	Variable	Yes	No
PENS[a]	Peak Electric	Yes	?[b]	No	Yes	Large	Yes	Storage
Iron	Iron	No	Yes	Yes	Yes	Large	Yes	No
Aircraft	H_2 Fuel	No	No	No	Yes	Medium	No	Yes
Hydrogen Direct Transport	H_2 Fuel	No	No	No	Yes	Small	No	Yes

[a]Peak electricity nuclear system. [b]Dependent upon specific facility and other factors

The world is rapidly exhausting its resources (Fig. 1) of the light crude oils (*Nature* 2004) used to make liquid fuels. These crude oils are the easiest to recover, have the lowest costs of recovery, and have the highest market prices. Consequently, for a century, oil companies have preferentially explored for and recovered these crude oils. Liquid fuels will increasingly be produced from other feedstocks with lower hydrogen-to-carbon ratios. In a refinery, these lower-grade feeds are converted to liquid fuels by adjusting the hydrogen-to-carbon ratio of the feedstock. For example, consider a heavy crude oil with a hydrogen-to-carbon ratio of slightly above 1. The hydrogen-to-carbon ratio of this feed can be converted to a liquid fuel by adding hydrogen (hydrocracking) or removing carbon (thermal cracking). The carbon that is removed is ultimately released to the atmosphere as carbon dioxide. Today, refineries use varying amounts of hydrocracking and thermal cracking for liquid-fuel production, depending upon hydrogen costs. If hydrogen is inexpensive, hydrogen will be added. If hydrogen is expensive, large quantities of carbon dioxide may be produced per unit of liquid fuel.

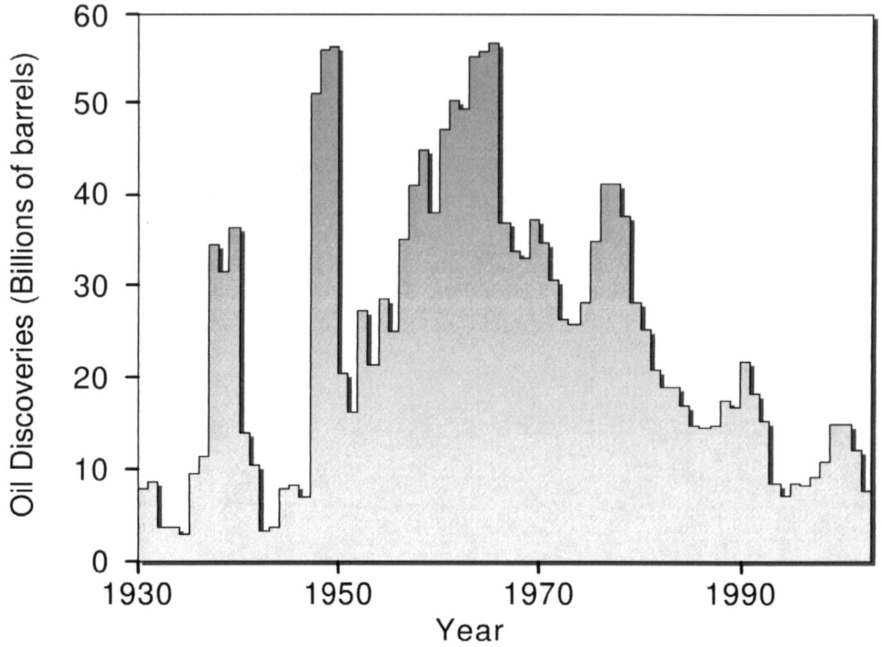

Fig. 1. Rate of discovery of conventional crude oils vs time.

If current technologies are used, the carbon dioxide emissions per unit of liquid-fuel production will likely increase dramatically in the next several decades as the hydrogen-to-carbon ratio of the feedstocks decreases (Fig. 2). Ultimately, in a business-as-usual scenario, the carbon dioxide emissions from the facilities producing the liquid fuels may exceed the carbon dioxide emissions from the actual burning of the liquid fuels in transport vehicles. *Alternatively, if economic hydrogen is available from non-greenhouse-emitting sources and the energy for the fuel processing does not release greenhouse gases to the atmosphere, the atmospheric carbon dioxide emissions from liquid-fuel production per unit of liquid fuel and the net atmospheric carbon dioxide emissions from the vehicles burning liquid fuels can drop dramatically—in principle, to zero.* In the extreme cases, the demand for hydrogen for liquid-fuel production approximately equals the energy content of all the liquid fuels we use today. The hydrogen demand as a function of feedstock is described herein.

3.1.1 Refineries

Typical refineries that convert light crude oil to liquid fuels are hydrogen neutral, neither consuming nor producing hydrogen. However, if low-cost hydrogen were available or there were penalties for carbon dioxide releases to the environment, the hydrogen-to-carbon ratio of fuels could be increased toward a maximum hydrogen-to-carbon ratio of ~2. In effect, the quantity (energy content) of liquid fuels produced per unit of crude oil or the equivalent transportation service provided per unit of crude oil is increased by 10 to 20%.

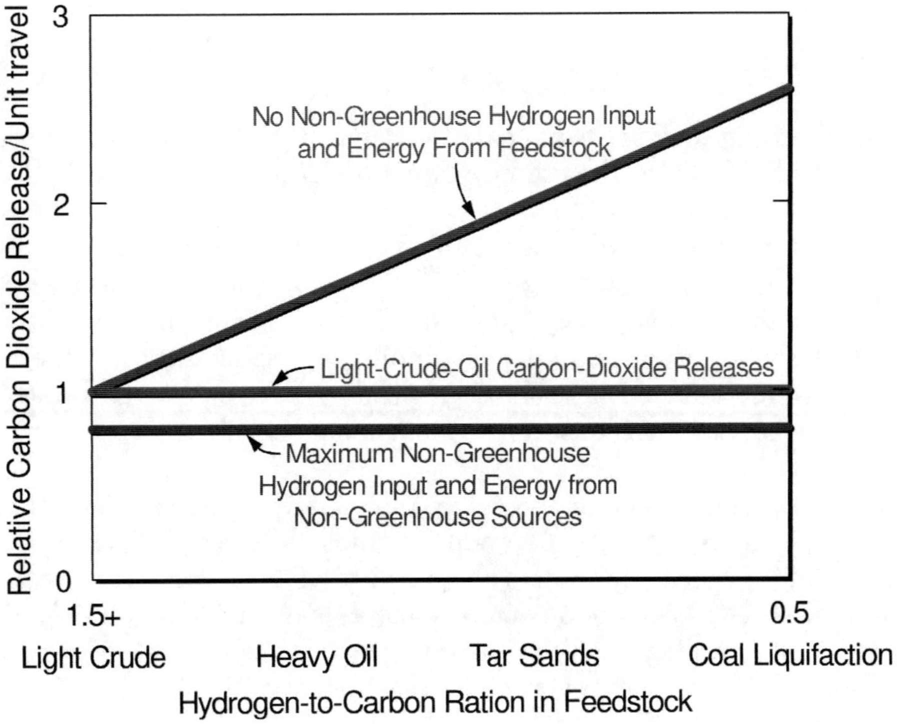

Fig. 2. Liquid-fuel carbon-dioxide production per vehicle mile vs feedstock, hydrogen source, and process energy source.

3.1.2 Heavy Oil and Tar Sands

Many experts believe that conventional oil production will peak in this decade and then decrease. The shortfalls in production are likely to be initially offset by conversion of heavy oils, tar sands, and other low-grade hydrocarbon deposits to liquid fuels (Williams 2003). The amount of synthetic crude oil that can be produced from these low-grade deposits far exceeds those of all conventional oil deposits combined. Massive amounts of hydrogen are required to convert these semiliquid, semisolid materials into gasoline. For example, the hydrogen-to-carbon ratio of tar sands is about 1, a value that must be raised to between 1.5 and 2 to convert the tar sands to gasoline.

Because of the imminent decline in conventional crude oils, the major oil companies of the world are rapidly expanding their facilities that convert these low-grade reserves into oil. Some perspective on the scale of operations can be obtained by examining the Alberta, Canada, tar sands developments. Production is being raised from its current level of 500,000 barrels/day of synthetic crude oil to 2.5 million barrels/day by 2010. Since 1996, $23 billion has been invested to increase production. An additional $37 billion in new plants and expansions has been announced. Most of the new world-class hydrogen plants are being built to support these facilities. If these tar-sand deposits are fully developed and natural gas is used to produce the required heat and hydrogen, the natural gas requirements will be *2 to 3 times the total projected Canadian natural gas reserves.* The demand of heavy oil and tar sands facilities represents the primary factor in the current growth in demand for hydrogen. This demand for hydrogen, and the resultant demand for natural gas to make the hydrogen, is beginning to have a serious impact on North American natural gas demand and prices.

3.1.3 Coal Liquefaction

The U.S. has some of the largest deposits of coal. Consequently, coal liquefaction is potentially a large long-term market for hydrogen to produce liquid fuels and minimize imports of oil.

In World War II, the Germans built large-scale facilities to convert coal to liquid fuels to replace imports of fuel. In 1955, South Africa started the first Sasol synthetic fuel plants that convert coal to liquid fuels using the German Fisher-Tropsch process. This complex was ultimately expanded to produce 100,000 barrels of oil per day. Following South Africa, in the 1980s New Zealand built a synthetic fuels plant using natural gas feed. This plant uses a two-step process that first produces methanol and then converts the methanol to gasoline.

There are two classes of coal liquefaction processes: direct and indirect. The direct processes (Beychok 1979) hydrogenate coal directly to produce liquids. The best of these processes have an efficiency of about 65%, which implies that about a third of the energy in the coal (and effectively one-third of the coal) is used to overcome these unavoidable process inefficiencies. Depending upon the coal, some faction of the remaining coal is used to make hydrogen; thus, perhaps a third of the carbon in the original coal is a component of the final liquid fuel.

The indirect processes convert carbon, oxygen, and water into synthesis gas—a mixture of hydrogen, CO, and CO_2. The newer synthetic fuel processes convert the synthesis gas into methanol (CH_3OH), which in turn is converted to liquid fuels. There are many variants of these processes. All existing commercial plants have used the indirect processes for a variety of reasons: the output of indirect processes is insensitive to the feedstock, the capital costs per unit of production are lower, the processes are technologically easier, and the products contain higher fractions of high-value liquid fuels. The process efficiency (Maiden 1988) for the New Zealand plant is estimated at 54% (ratio of energy value of liquid fuels to energy input).

These processes can be considered black boxes, carbon, oxygen, and water go into the process and liquid fuels and carbon dioxide exit the process. The carbon is used (1) as a chemical feedstock to provide the carbon in the liquid fuel and (2) as a fuel to provide energy for the process—including hydrogen production. At a fundamental level, the feed can be any carbon-containing material: coal, garbage, natural gas, etc.

The economics of liquid-fuel production require very large plants with massive demands for hydrogen and, in most cases, massive quantities of oxygen. The large German program in the 1970s to produce hydrogen using nuclear energy was to supply hydrogen and energy for coal liquefaction. (Germany has large coal resources but no liquid fuels.) The smaller nuclear hydrogen program in the United States in the 1970s had the same goal.

3.1.4 Direct Atmospheric Fuel Production

Liquid fuels can be made from water and carbon dioxide extracted from (1) the atmosphere or (2) the ocean. In this case, external sources of energy must be used. This energy can be in the form of electricity and hydrogen, where the hydrogen is used (1) as a feedstock to make the liquid fuels and (2) as an internal energy source to drive the process of producing the fuel. Given unlimited hydrogen, this option provides unlimited liquid fuel. No greenhouse impacts occur because the carbon dioxide is recycled from the atmosphere or seawater. There have been a series of studies on the technology for liquid-fuel production and the cost implications.

- *Navy.* The primary logistical demand for the Navy is the provision of liquid fuels for aircraft and ships. Aircraft carrier flight operations and fleet operations are limited by the capability of the oil tankers to provide fuel. This constraint can be eliminated by the use of a nuclear-powered tanker (Terry 1995) that manufactures jet fuel with nuclear hydrogen from seawater and carbon dioxide from the air or ocean.

- *Fusion energy complexes.* Engineering studies indicate that likely fusion energy plants may be very large [several thousand megawatts (electric) equivalent], even relative to the size of fission nuclear power plants. The very large energy output of these machines would create major difficulties in electrical transmission. These constraints would be eliminated if there were a large local market for that energy such as the production of liquid fuels (Fillo, Powell, and Steinberg, 1981). A series of studies were done to evaluate alternative liquid fuels production with the carbon from atmospheric carbon dioxide.

Most of these studies have assumed that the hydrogen is produced through electrolysis. Detailed studies (Steinberg and Dang, 1975) have examined alternative methods for recovery of carbon dioxide from the atmosphere and ocean. The generally recommended atmospheric-carbon-dioxide-recovery method is the use of aqueous potassium carbonate (K_2CO_3) solution for absorption and stripping of atmospheric carbon dioxide. About 80% of the total energy input required to produce the liquid fuel is used in the processes to produce hydrogen from water. The hydrogen is then used to produce the liquid fuel. If a high-temperature reactor is assumed to have 50% efficiency in the conversion of heat to electricity with hydrogen made by electrolysis, about 30% of the thermal energy produced by a nuclear reactor is converted into liquid fuels. This technology has several implications:

- *Liquid-fuel costs.* This option provides unlimited liquid fuels with no greenhouse impacts as long as the hydrogen and electricity come from non-greenhouse energy sources. If there is a non-greenhouse source of energy, it caps the potential costs of liquid fuels because the raw materials (water and atmosphere) are available in unlimited quantities.

- *Hydrogen economy.* From an economic perspective, this technology places an upper economic limit on the allowable costs for using hydrogen directly as a fuel compared with using liquid fuels—if there are constraints on net release of carbon dioxide to the atmosphere. About 80% of the energy for liquid-fuel production using carbon dioxide from the atmosphere or ocean is the cost of hydrogen manufacture. The other energy costs are associated with capturing carbon dioxide from the atmosphere and converting the carbon and hydrogen to liquid fuels. If the costs to distribute and use hydrogen directly as a fuel exceed those of liquid fuels, this technology defines an economic upper limit for the direct use of hydrogen for transportation before it becomes more economic to make liquid fuels from hydrogen and atmospheric carbon dioxide.

3.1.5 Biomass Conversion

Biomass today is used to produce liquid fuels such as alcohol by fermentation. This is potentially a non-greenhouse liquid-fuel source because the carbon dioxide used to make the biomass comes from the atmosphere. However, in a plant that converts biomass to liquid fuel, part of the carbon in the feed is converted to carbon dioxide by the biomass to provide the energy to keep the biomass alive. For example, the conversion of corn to ethanol results in roughly a third of the carbon from the original corn in the ethanol, one-third in the by-product animal feed, and one-third in the carbon dioxide released to the atmosphere. Biomass processing facilities are concentrated sources of carbon dioxide recovered from the atmosphere.

This carbon dioxide can be combined with hydrogen to produce liquid fuels. In effect, there is the potential to use biomass as a non-greenhouse-generating carbon source to make liquid fuels as well as an energy source. This type of option has been investigated by Bruce Energy Center of Ontario Hydro and Integrated Energy Development Corporation. The concept was to produce hydrogen through electrolysis of water using electricity from the Bruce Nuclear Generating Station in Tiverton, Ontario, Canada, and to collect carbon dioxide from an ethanol plant. The feeds would then be converted into liquid fuels and chemicals using existing technology (Gurbin and Talbot 1994). This type of liquids fuel production is distributed because of the cost to collect and transport biomass to a central fuel plant location.

3.1.6 Carbon-Dioxide Vehicle Recycle Systems

Liquid fuel systems that deliver hydrogen to the vehicle engines, capture the carbon dioxide on-board the vehicle, and recycle carbon dioxide, are being investigated in Japan (Kato et al., 2003). The vehicle is fueled with a liquid fuel and a calcium oxide (CaO) bed. The system contains the following components.

- *Vehicle steam reformer*. The CaO bed is used as a steam reformer where the liquid fuel is converted to hydrogen and carbon dioxide and the carbon dioxide reacts with the CaO to form solid calcium carbonate ($CaCO_3$). The reaction of the CaO and the CO_2 (1) is highly exothermic and provides the energy necessary to drive the highly endothermic stream reforming reaction to completion to maximize hydrogen production and (2) removes all the CO_2 and thus drives the equilibrium reactions to produce hydrogen rather than a mixture of H_2, CO, and CO_2.

- *Vehicle engine*. Fuel cells or an internal combustion engine powers the vehicle with hydrogen.

- *Fuel factory*. The $CaCO_3$ bed is returned to a fuel factory where hydrogen chemically reduces the $CaCO_3$ to CaO for recycle to vehicles and the recovered carbon dioxide is combined with hydrogen to produce new liquid fuel.

In this system, carbon is a recyclable hydrogen and energy storage mechanism between the fuel factory and the vehicle. Hydrogen enters the fuel factory and reappears inside the vehicle. Current estimates indicate that the volume and mass of this hydrogen fuel delivery system on-board the vehicle are less than other existing methods to deliver hydrogen to a vehicle. The low volume and mass of the fuel system aboard the vehicle is because energy is stored in two high energy-density forms: (1) the solid CaO reformer beds and (2) the liquid fuel. Most fuel factories would be near cities to minimize transport cost of solid CaO reformer beds. This system configuration implies distributed fuel factories requiring hydrogen—probably via pipeline.

3.2 Ammonia and Other Chemical Applications

Ammonia production (fertilizer) consumes about half the hydrogen produced today and is the primary chemical industry use of hydrogen. The market is growing slowly and is international. Ammonia is made where there is inexpensive natural gas that provides inexpensive hydrogen and shipped to the customer. Because of the increasing use of precision agriculture that has lowered the nitrogen fertilizer inputs per unit of food or biomass produced, large growth in demand for ammonia (and thus hydrogen) is not expected in this industry.

3.3 Energy Storage for Intermediate and Peak Electrical Production

The demand for electricity varies daily, weekly, and seasonally. As a result, the market price of electricity varies by an order of magnitude as a function of time. To meet this demand, the utilities buy lower-capital-cost peaking power units, typically gas turbines that burn natural gas. In addition, utilities have developed storage devices so that they can buy electricity during times of low demand and low cost and sell the electricity during times of high demand and high prices.

For example, the Tennessee Valley Authority (TVA) operates the Raccoon Mountain pumped-hydro storage facility. This facility pumps water up the mountain when low-cost power is available. During times of high power demand and high-priced electricity, the water direction is reversed to produce electricity. This facility has a rated capacity of 1530 MW(e). Another example is the Duke Power Koewee-Jocasse Project where water is pumped from Lake Koewee to Lake Jocasse during times of low power demand and the water is released through turbines to generate electricity during times of peak power demand.

The variability of the price of electricity creates the potential for a large hydrogen market aimed at producing electrical power at those times of day when the price of electricity is at its maximum. A *peak electricity nuclear system* (PENS) using nuclear hydrogen (Forsberg 2005) has been proposed that consists of three components (Fig. 3):

- *Hydrogen production.* A nuclear power plant with an associated hydrogen and oxygen (optional) production plant to produce hydrogen and oxygen at a constant rate to minimize the production costs.

- *Hydrogen storage.* One or more underground facilities for the low-cost storage of hydrogen and oxygen (optional). Underground caverns, depleted oil and gas fields, and aquifers are the traditional approaches to the low-cost storage of natural gas to meet seasonally variable natural gas demand. About 8 trillion cubic feet of natural gas can be stored in existing underground facilities in the United States (half of this quantity as a buffer gas to maintain high pipeline pressures). In countries such as Great Britain, salt caverns have been used for many decades for the low-cost storage of hydrogen. Underground storage is the only known low-cost technology for storing compressed gases; however, the economics demand very large facilities.

- *Peak electric production.* Large banks of fuel cells convert hydrogen to electricity during periods of high demand for electrical power and associated high prices for electricity. For every megawatt of steady-state hydrogen production from the nuclear reactor, there would be several megawatts of fuel cells. While the reactor produces hydrogen at a constant rate, the fuel-cell electrical production is highly variable—from zero to many times the rate of energy production from the reactor when the price of electricity is high. The fuel cells may be placed at the reactor site. As an alternative, to reduce electrical grid requirements, hydrogen pipelines may be built around major cities, with large fuel cell facilities located at junctions where the pipeline crosses long-distance transmission lines.

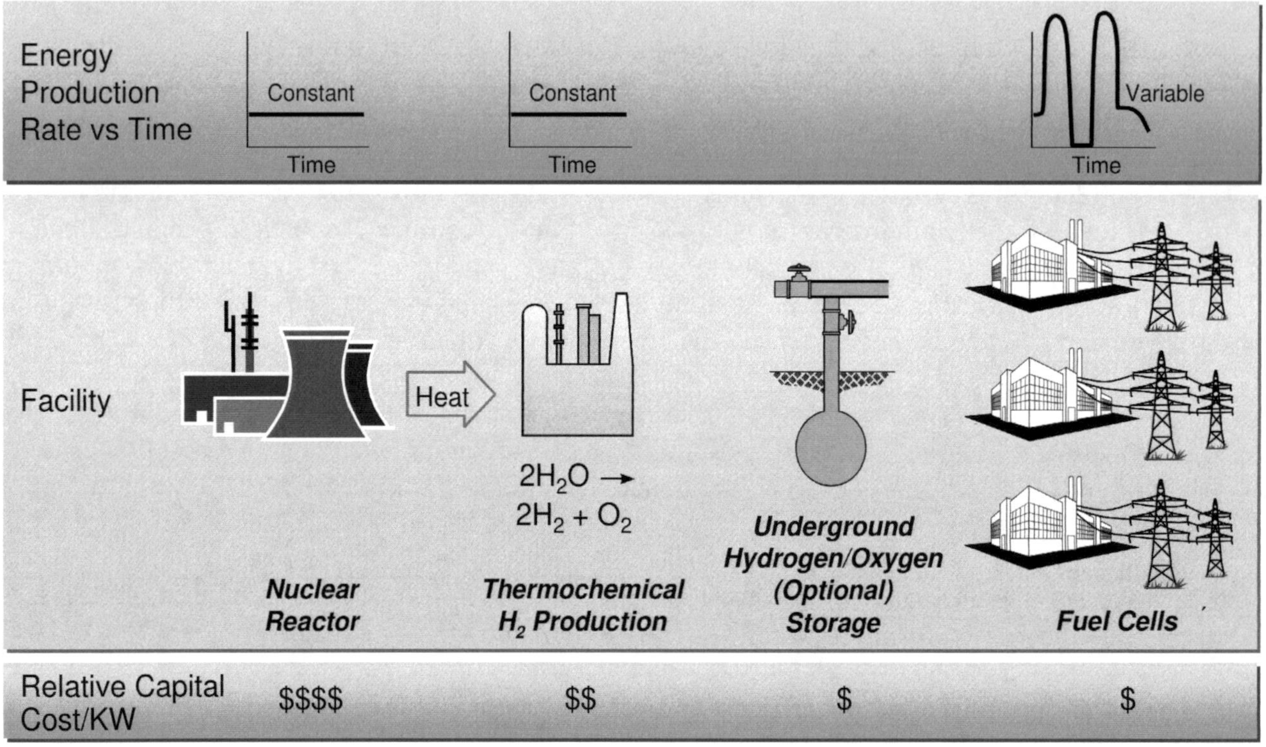

Fig. 3. Peak Electricity Nuclear System using nuclear hydrogen.

The economic feasibility of PENS is based on projected costs and efficiencies of fuel cells. The goal of fuel cell developers is to reduce the capital cost for hydrogen fuel cells to <$100/kW(e) with efficiencies of about 70%, compared with costs of $500/kWe for gas-turbine plants burning natural gas with efficiencies of about 50%. This use of fuel cells allows PENS to have a very large capacity to produce peak electrical power compared with the steady-state output of the reactor. The capital costs of fuel cells per kilowatt electric may be further reduced by the use of oxygen rather than air. Hydrogen is produced at a constant rate to minimize hydrogen production costs. However, but the ultimate product, electricity, is sold at times of peak demand for premium prices.

PENS is a stand-alone facility, like the TVA Raccoon Mountain project. It is not dependent on the outside market for hydrogen. Should another demand for hydrogen exist, however, the hydrogen can be sold at the plant gate. In the United States, the potential market exceeds 100 GW(e) equivalent on a steady-state basis if the peak and much of the intermediate load are replaced by PENS. There are several variants of this concept using various hydrogen production and electricity production technologies—all dependent upon the low-cost storage of hydrogen and oxygen on a large scale.

3.6 Direct Reduction of Iron Ore

In the production processes for converting iron ores into iron and steel, carbon, primarily in the form of coke, has been traditionally used to reduce the iron oxides to iron metal. However, in the last several decades, there has been increasing production of iron using the direct reduction iron (DRI) process. In 1998, about 4% of the primary iron in the world was produced by the DRI process with rapid growth in iron production. In the DRI process, syngas (a mixture of hydrogen and carbon monoxide) made from natural gas is used to reduce iron ores to iron. The major chemical reactions are as follows:

$$Fe_3O_4 + CO \rightarrow 3FeO + CO_2$$

$$Fe_3O_4 + H_2 \rightarrow 3FeO + H_2O$$

$$FeO + CO \rightarrow Fe + CO_2$$

$$FeO + H_2 \rightarrow Fe + H_2O$$

The DRI process has lower capital costs than alternative methods used to produce iron but requires a low-cost source of hydrogen. The primary market for DRI is to provide a purified iron feed for electric arc furnaces (EAFs) that produce various steel products. EAFs have lower capital costs than traditional steel mills and are environmentally cleaner operations than blast furnaces. Over a third of the world's steel production uses this process. It is predicted that by 2010 up to 45% of the world's steel may be made with EAFs. Historically, scrap metal has been the traditional feed for EAFs. However, there are two constraints: the availability of scrap metal and the various difficult-to-remove impurities (copper, nickel, chrome, molybdenum, etc.) that are present in the lower-grade scrap metal. Blending clean DRI-process iron with scrap metal dilutes the impurities below the level that affect product quality.

Iron production is a significant existing market for hydrogen. If low-cost hydrogen were available, the DRI process would replace other methods of iron production. The economics of DRI relative to other processes depend upon three factors.

- *Technological developments.* The continuing improvements in EAF technology in terms of reduced production costs and increased capabilities to produce higher-quality steel have expanded the market share of this technology. That creates the demand for more iron by the DRI process as traditional sources of scrap metal are exhausted.

- *Environmental protection.* Traditional steel processes use coal and generate large quantities of pollutants. Clean air requirements strongly affect the economics of these competing processes.

- *Hydrogen costs.* The process is used where there is low-cost natural gas for hydrogen production near iron deposits.

A variant of this option was studied in Japan between 1973 and 1980. The "Nuclear Steelmaking Project" conducted major engineering tests and designed a 500-MW(t) reactor. This project was the start of the Japanese high-temperature reactor program.

3.7 Aircraft

Liquid hydrogen has been considered for three types of aircraft applications: commercial jets, hypersonic jets, and electric airplanes.

Commercial jets. Commercial aircraft can be fueled with liquid hydrogen. The European Union (Airbus Deutschland GmbH 2003) funded a consortium of 35 partners from the aviation sector, led by Airbus Deutschland, to conduct a systems analysis of hydrogen-fueled aircraft—the CRYOPLANE project. This consortium examined a wide range of aircraft from business jets to large long-range aircraft such as the new jumbo Airbus A380. The key issue was to model the liquid-hydrogen fuel system. Per unit energy, liquid hydrogen has four times the volume of jet fuel; therefore, the fuel tanks must be four times larger. Analysis showed that because of the larger external surface area of the aircraft needed to accommodate the fuel tanks, the energy consumption would increase by 9 to 14%. Overall operating costs would increase by 4 to 5% based on fuel alone. It was also concluded that the engines would be equally efficient, the aircraft would have safety equivalent to that of current aircraft, and the environmental impacts would be substantially less (i.e., no carbon dioxide emissions). Further development is needed; however, such an aircraft system could be implemented within 15 to 20 years of a decision to use hydrogen as a fuel.

Hypersonic jets. Hypersonic aircraft require liquid-hydrogen fuel. The air velocity through a ram jet requires a fuel with a very fast flame temperature and diffusion rate to ensure combustion within the engine. Hydrogen is the only option. Liquid hydrogen is also used to provide active cooling to leading aircraft surfaces to avoid melting. This market depends upon the development of these jets.

Electric aircraft. Aircraft can use electric motors and hydrogen fuel cells to power propellers. These systems have potentially very high efficiencies compared with alternative propulsion systems. If weight and cost can be sufficiently reduced, liquid hydrogen becomes a preferred fuel. The viability of such aircraft depends upon technological progress in fuel cells and related systems.

3.8 Hydrogen Auto and Truck Transport System

Hydrogen is proposed as the ultimate transport fuel for cars, trucks, and buses. Recent reports (U.S. National Research Council 2004) describe the various scenarios. This can be considered the ultimate end state of hydrogen development if the various technical barriers are eliminated. Some of the hydrogen transport scenarios assume on-board conversion of liquids to hydrogen. Those scenarios are similar to the liquid-fuel scenarios described earlier, except that there would be a strong emphasis to maximize the hydrogen content of the liquid fuel. Other scenarios assume hydrogen as the stored fuel.

4. Conclusions

The future will demand very large quantities of hydrogen; however, it is unclear whether that hydrogen will be used in vehicles or to produce a variety of liquid fuels at large industrial complexes. Constraints on the release of carbon dioxide to the environment would create massive added demands for hydrogen using technologies that do not release carbon dioxide to the environment. The hydrogen demand could equal 40% of the total energy demand if it replaces oil. Many pathways to a hydrogen future involve increasing uses of hydrogen for making liquid fuels from different feedstocks, meeting peak electrical power demands, and producing steel. As a consequence, the transition to a hydrogen economy may be facilitated if the first users are large industrial companies with sizable demands for hydrogen and minimal need for large supporting hydrogen infrastructures.

REFERENCES

Airbus Deutschland GmbH (2003), *Liquid Hydrogen Fueled Aircraft—Systems Analysis*, European Community Competitive and Sustainable Growth Subprogram (September 23).

Beychok, M. R. (1979), "Coal, Liquefaction" *in Encyclopedia of Chemical Processing and Design*, ed, J. J. McKetta, Marcel Dekker, Inc., New York.

Fillo, J. A., J. Powell, and M. Steinberg (1981), "Synthetic Fuels and Fusion," *Nucl. Engin. and Des.*, **63** (2), 331–335.

Forsberg, C. W. (January 2005), *Nuclear Hydrogen for Peak Electricity Production and Spinning Reserve*, ORNL/TM-2004/194, Oak Ridge National Laboratory, Oak Ridge, Tennessee (January).

Gurbin, G. M. and K. H. Talbot (1994), "Nuclear Hydrogen—Cogeneration and the Transition Pathway to Sustainable Development," *9th Pacific Basin Nuclear Conference*, *Trans. Am. Nucl. Soc.* **70** (Suppl. 1).

Herzog, H., and D. Golumb (2004), "Carbon Capture and Storage from Fossil Fuels," *in Encyclopedia of Energy,* 6 vols., ed. C. J. Cleveland, Elserier, Amsterdam.

Kato Y., K. Ando, and Y. Yoshizawa (2003), "Study of a Regenerative Fuel Reformer for a Zero-Emission Vehicle System," *J. of Chemical Engineering of Japan*, **36**, 860–866.

Maiden, C. J. (1988), "The New Zealand Gas to Gasoline Project," *Chemtech* **18** (1).

Nature (2004), **17**(6993), 694-695.

Steinberg, M. and V. D. Dang (1975), *Use of Controlled Thermonuclear Reactor Fusion Power for the Production of Synthetic Methanol Fuel from Air and Water*, BNL-20016, Brookhaven National Laboratory, Upton, New York (April).

Terry, K. B. (1995), *Synthetic Fuels for Naval Applications Produced Using Shipboard Nuclear Power*, Nuclear Engineer and Master of Science in Nuclear Engineering Thesis, Department of Nuclear Engineering, Massachusetts Institute of Technology, Cambridge, Massachusetts.

U.S. National Research Council (2004), *The Hydrogen Economy: Opportunities, Costs, Barriers, and R&D Needs*, National Academy Press, Washington, D.C.

Williams, B. (2003), "Heavy Hydrocarbons Playing Key Role in Peak-Oil Debate, Future Energy Supply," *Oil Gas J.* **101** (29), 20–27.

73f Economics of Using High Temperature Electrolysis and Advanced Nuclear Cycles for at Sea Synthesis of Logistic Fuel for the United States Navy

Charles Smith, Lyman J Frost, Robert C Carrington

The United States Navy is interested in achieving logistic fuel independence by using dedicated nuclear power to produce hydrocarbon fuels onboard naval vessels. The production of the fuels will involve use of electrical and thermal energy from a reactor that will be used to produce hydrogen from seawater. A carbon source will be provided by extracting carbon dioxide from either air, water, or the exhaust gases of non-nuclear escort vessels. Paragraph1Paragraph2

The technology exist today to syntesize fuels given sufficient energy. Hydrogen can be generated from seawater using conventional electrolysis or high temperature electrolysis. Carbon dioxide can be captured using technoloy such as amine scrubbing of stack gases, membrane separation from air, or vacuum treatment of seawater. With a source of hydrogen and carbon traditional methods such as Fischer Tropsch can be used to produce the desired fuels. Paragraph2Paragraph3

This paper will present results comparing the volumes and energy required to produce a nominal 100000 gallons (~700000 pounds) of JP5 under alternative scenarios. The paper will compare existing technologies with potential savings generated by advanced technologies such as high temperature electrolysis and an advanced high temperature nuclear cycle.

74a A Novel Method for Producing Hydrogen Based on the Ca-Br Cycle

Christopher D McGrady, Michael F Simpson, Prateek Sachdev, Vivek Utgikar

While the Ca-Br Cycle is a promising method for efficiently producing hydrogen from water, it suffers from limitations inherent to gas-solid reactions. The cycle depends on the repeatable transformation of calcium bromide to calcium oxide and back to calcium bromide. The use of pure solids for these reactions should lead to rapid solid particle degradation and slow reaction kinetics. To circumvent this problem, it has been proposed to run the reaction in a molten salt phase. A new reactor concept has, thus, been developed that includes molten calcium bromide with a yet to be determined concentration of calcium oxide. Preliminary experimental results indicate that the solubility of calcium oxide in calcium bromide at 800 C is at least 1.2 wt%, thus indicating that this is still a promising concept. Early attempts to hydrolyze molten calcium bromide indicated that solid calcium oxide formation may inhibit reactivity, and thus injection of the moisture into the salt must be optimized.

74b Hydrogen from the Calcium-Bromine Cycle - Laboratory Demonstration Requirements

Diana T Matonis, Richard D Doctor

In developing the Ca-Br thermochemical cycle for nuclear hydrogen production a special focus is to determine the feasibility of developing a continuous process. Water-splitting with HBr formation takes place at a temperature of 1030 K when steam reacts with CaBr2. The issue is now to find materials compatible with molten CaBr2 so that the heat of fusion from freezing the CaBr2 can supply the needed heat of reaction. The HBr then formed used a cold plasma dissociation for H2 generation; this was followed by a molten spray contactor reactors for CaBr2 regeneration. The most promising approach for the Ca-Br thermochemical cycle was identified and flowsheet analysis was performed to estimate performance and efficiency. This task developed a conceptual Ca-Br system design necessary to identify key materials and scale-up issues.

74c Computational Fluid Dynamic (CFD) Modeling of Bipolar Steam Solid Oxide Electrolyzer Cells (SOECs)

Diana T Matonis, Richard D Doctor

The electrolysis cell can be viewed as a chemical reactor dissociating steam and requiring the input of electricity. In its planar configuration, the planar solid oxide electrolysis cell (PSOEC) with reagent water coming in as high-temperature steam can be either electrolyte or electrode supported. The computational fluid dynamics (CFD) modeling analysis in this project focuses on a single planar PSOEC since its fabrication cost is estimated to be the lowest because of ease of manufacture. This is the configuration which Idaho National Laboratory and Ceramatec, Inc., with whom we collaborate, will pursue experimentally. Heat transfer and electrochemical efficiencies will degrade depending upon the extent of flow field deviations from uniformity inside and outside each cell comprising a stack. Therefore, to achieve the highest net efficiency and the lowest net cost, a detailed understanding of the thermal and flow processes both inside and the balance of plant outside of a SOEC is required. This will then be used to improve overall system performance.

74d Electrochemical Generation of Hydrogen via Thermochemical Cycles

PremKumar Sivasubramanian

Reforming carbon-based fuels (e.g., natural gas) to produce hydrogen results in impurities (e.g., CO), which can poison fuel-cell catalysts. Water electrolysis does produce highly pure hydrogen, but it consumes a substantial amount of electrical energy. Alternatively, a variety of thermochemical cycles could be used to produce pure hydrogen in an energy efficient manner. Two promising cycles which are of interest are based on the internal recycle of sulfuric acid and bromine. We present here, an electrochemical process that produces pure hydrogen and recycles sulfuric acid and bromine via the electrochemical oxidation of SO2 and HBr, respectively. Both of these electrochemical reactions take place in a proton exchange membrane (PEM) reactor, similar in construction to a PEM fuel cell. The reactants, either SO2 or HBr, are fed to the anode and water is fed to cathode. Protons are formed at the anode, and transported across a Nafion membrane, which are then reduced at the cathode to form hydrogen. We will show that the performance of the electrolytic reactor (i.e., current versus voltage) is a strong function of reactor design and operating conditions. This performance will dictate the size and energy consumption of a hydrogen production process.

PROGRESS IN HIGH-TEMPERATURE ELECTROLYSIS FOR HYDROGEN PRODUCTION USING PLANAR SOFC TECHNOLOGY

J. Stephen Herring, James E. O'Brien, Carl M. Stoots, G. L. Hawkes
Idaho National Laboratory
2525 N. Fremont Ave., Idaho Falls, ID, 83415, USA

Joseph J. Hartvigsen
Ceramatec, Inc.
2425 South 900 West, Salt Lake City, UT 84119, USA

Mehrdad Shahnam
FLUENT Inc.
3647 Collins Ferry Road, Suite A, Morgantown, WV, 26505, USA

ABSTRACT

A research program is under way at the Idaho National Laboratory to assess the performance of solid-oxide cells operating in the steam electrolysis mode for hydrogen production over a temperature range of 800 to 900ºC. The research program includes both experimental and modeling activities. Selected results from both activities are presented in this paper. Experimental results were obtained from a ten-cell planar electrolysis stack, fabricated by Ceramatec[1], Inc. The electrolysis cells are electrolyte-supported, with scandia-stabilized zirconia electrolytes (~140 μm thick), nickel-cermet steam/hydrogen electrodes, and manganite air-side electrodes. The metallic interconnect plates are fabricated from ferritic stainless steel. The experiments were performed over a range of steam inlet mole fractions (0.1 – 0.6), gas flow rates (1000 – 4000 sccm), and current densities (0 to 0.38 A/cm^2). Hydrogen production rates up to 90 Normal liters per hour were demonstrated. Stack performance is shown to be dependent on inlet steam flow rate. A three-dimensional computational fluid dynamics (CFD) model was also created to model high-temperature steam electrolysis in a planar solid oxide electrolysis cell (SOEC). The model represents a single cell as it would exist in the experimental electrolysis stack. Mass, momentum, energy, and species conservation and transport are provided via the core features of the commercial CFD code FLUENT[1]. A solid-oxide fuel cell (SOFC) model adds the electrochemical reactions and loss mechanisms and computation of the electric field throughout the cell. The FLUENT SOFC user-defined subroutine was modified for this work to allow for operation in the SOEC mode. Model results provide detailed profiles of temperature, Nernst potential, operating potential, anode-side gas composition, cathode-side gas composition, current density and hydrogen production over a range of stack operating conditions. Mean model results are shown to compare favorably with the experimental results obtained from the ten-cell stack tested at INL.

INTRODUCTION

Currently there is strong interest in the large-scale production of hydrogen as a secondary energy carrier for the non-electrical market. Hydrogen is of particular interest as a

[1] References herein to any specific commercial product, process, or service by trade name, trademark, manufacturer, or otherwise, does not necessarily constitute or imply its endorsement, recommendation, or favoring by the U.S. Government, any agency thereof, or any company affiliated with the Idaho National Engineering and Environmental Laboratory

secondary energy carrier because it has the potential to be storable, transportable, and environmentally benign. Hydrogen is now produced primarily via steam reforming of methane. From a long-term perspective, methane reforming is not a viable process for large-scale production of hydrogen as a major energy carrier since such fossil fuel conversion processes consume non-renewable resources and emit greenhouse gases to the environment. Consequently, there is a high level of interest in production of hydrogen from water splitting via either thermochemical or electrolytic processes (e.g., National Research Council, 2004).

High-temperature nuclear reactors have the potential for substantially increasing the efficiency of hydrogen production from water splitting, with no consumption of fossil fuels, no production of greenhouse gases, and no other forms of air pollution. Thermal water-splitting for hydrogen production can be accomplished via high-temperature electrolysis or thermochemical processes, using high-temperature nuclear process heat. In order to achieve competitive efficiencies, both processes require high-temperature operation (~850°C). Thus these hydrogen-production technologies are tied to the development of advanced high-temperature nuclear reactors. High-temperature electrolytic water splitting supported by nuclear process heat and electricity has the potential to produce hydrogen with overall system efficiencies near those of the thermochemical processes (Yildiz, 2003; IAEA, 1999), but without the corrosive conditions of thermochemical processes and without the fossil fuel consumption and greenhouse gas emissions associated with hydrocarbon processes. Specifically, a high-temperature advanced nuclear reactor coupled with a high-efficiency high-temperature electrolyzer could achieve a competitive thermal-to-hydrogen conversion efficiency of 45 to 55%.

A research program is under way at the Idaho National Laboratory (INL) to simultaneously address the technical and scale-up issues associated with the implementation of solid-oxide electrolysis cell technology for hydrogen production from steam. The research program includes an experimental program aimed at performance characterization of electrolysis cells and stacks. Results of single-cell tests have been documented in several recent papers (O'Brien et al., Herring et al., 2004). Single (button) cell tests are useful for basic performance characterization of electrode and electrolyte materials and of different cell designs (e.g., electrode-supported, porous metal substrate-supported). The single-cell results demonstrated efficient small-scale hydrogen production, with performance close to theoretical predictions. Cell performance was shown to be continuous from the fuel-cell to the electrolysis mode. The effects of steam starvation and thermal cycling on cell performance parameters were demonstrated. Based on these preliminary results, high-temperature electrolysis appears to be a viable means for hydrogen production using nuclear energy.

The INL research program includes experimental, modeling and materials-development tasks. The experimental results presented in this paper were obtained from a ten-cell planar electrolysis stack fabricated by Ceramatec, Inc.. Experimental results include voltage-current behavior, stack area-specific resistance, hydrogen production rates, electrolysis efficiency, and internal stack temperature measurements over a range of operating conditions. For detailed SOEC modeling, the commercial CFD code FLUENT was selected. Fluent Inc. was funded by the US Department of Energy National Energy Technology Laboratory (DOE-NETL) to develop a solid-oxide fuel cell (SOFC) module for coupling to the core mass, momentum, energy, and species conservation and transport features of the FLUENT CFD code (Prinkey et al., 2004). The SOFC module adds the electrochemical

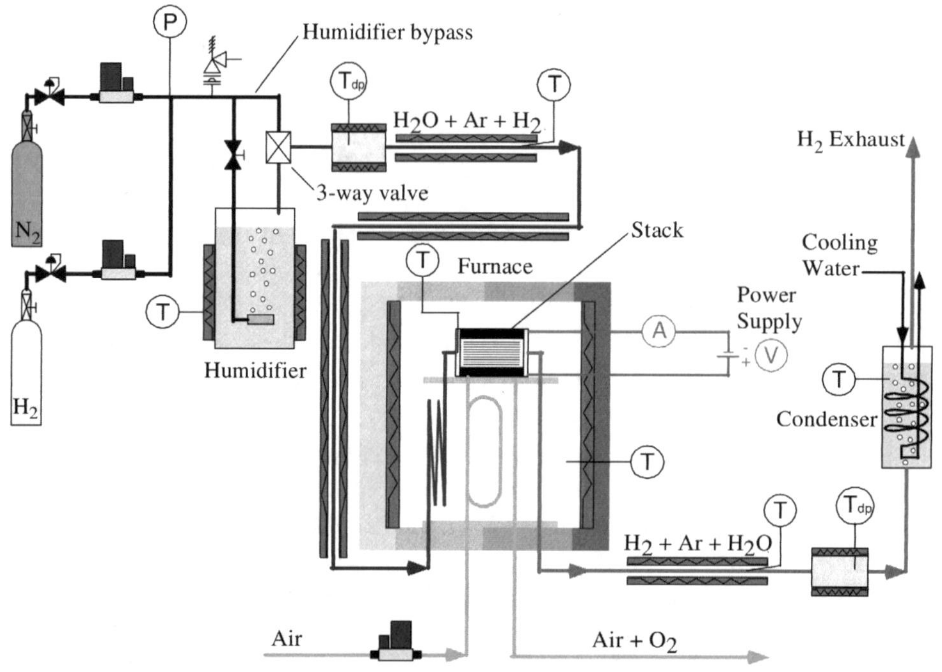

Figure 1: Schematic of experimental apparatus for electrolysis stack testing.

reactions and loss mechanisms and computation of the electric field throughout the cell. The FLUENT SOFC user-defined subroutine was modified for this work to allow for operation in the SOEC mode. Model results provide detailed profiles of temperature, Nernst potential, operating potential, anode-side gas composition, cathode-side gas composition, current density and hydrogen production over a range of stack operating conditions. Results of the numerical model are compared to experimental results obtained from a ten-cell stack tested at INL.

APPARATUS

A schematic of the stack-testing apparatus is presented in Fig. 1. Primary system components include gas supply cylinders, gas mass-flow controllers, a humidifier, dewpoint measurement stations, temperature and pressure measurement, high temperature furnace, and a solid oxide electrolysis stack. Nitrogen was used as an inert carrier gas. The use of a carrier gas allows us to independently vary both the partial pressures and the flow rates of the steam and hydrogen gases while continuing to operate at atmospheric pressure. The flow rates of nitrogen, hydrogen and air are established by means of precision mass-flow controllers (Hastings Model HFC-302, with Hastings Model 400 electronics). Air flow to the stack is supplied by the laboratory shop air system, after passing through a two-stage extractor / dryer unit.

Downstream of the mass-flow controller, nitrogen is mixed with a smaller flow of hydrogen gas. Hydrogen is included in the inlet flow as a reducing gas in order to prevent oxidation of the nickel cermet electrode material. The nitrogen / hydrogen gas mixture is mixed with steam by means of a heated humidifier. The humidifier consists of a heated stainless-steel vessel containing demineralized / deionized water through which the nitrogen / hydrogen flow is bubbled. The dewpoint temperature of the steam / hydrogen / nitrogen gas mixture exiting the humidifier is monitored continuously using a precision dewpoint sensor (Vaisala Model HMP247). The humidifier is fitted with a clamp heater and is externally insulated. The humidifier temperature is maintained at a constant setpoint value using feedback control. Since the vapor pressure of the water and the resulting partial pressure of the steam exiting the humidifier are determined by the water bath temperature, the water vapor mass flow rate is directly proportional to the carrier gas flow rate for a specified bath temperature. Also, since the nitrogen and hydrogen flow rates are fixed by the mass flow controllers, and the steam partial pressure is fixed by the bath temperature, the complete gas composition is precisely known at all times. All gas lines located downstream of the humidifier are heat-traced in order to prevent steam condensation. Gas line temperatures are monitored by thermocouples and controlled by means of variable transformers.

Figure 2. Detail of stack.

A close-up photograph of the test article, a 10-cell solid-oxide electrolysis stack, is shown in Fig. 2. The stack was fabricated by Ceramatec, Inc. of Salt Lake City, UT. The per-cell active area is 64 cm^2. It is designed to operate in cross flow, with the steam / hydrogen gas mixture entering the inlet manifold on the right in the photograph, and exiting through the outlet manifold, visible on the left in the photograph. Air flow enters at the rear though an air inlet manifold (not visible in Fig. 3) and exits at the front directly into the furnace. The power lead attachment tabs, integral with the upper and lower interconnect separator plates are also visible in the photograph. Stack operating voltages were measured using wires that were directly spot-welded onto these tabs. Since the stack air outlet plane is not enclosed, the small air flow channels are accessible for instrumentation. Four intermediate cell voltages were monitored using small-diameter wires inserted into these air flow channels. In addition, four miniature thermocouples (inconel-sheathed, 0.010-inch (250 μm) OD, mineral-insulated, ungrounded, type-K) were inserted into the air flow channels to monitor internal stack temperatures.

The internal components of the stack are shown in Fig. 3. The interconnect plate, shown on the left in Fig. 3, is fabricated primarily from low-chromium ferritic stainless steel. It includes an impermeable separator plate (~0.46 mm thick) with edge rails and two corrugated / perforated "flow fields," one on the air side and one on the steam/hydrogen side. The height of the flow channel formed by the edge rails and flow fields is 1.0 mm. Each flow field includes 32 perforated flow channels across its width to provide uniform gas-flow distribution. The steam / hydrogen flow field (shown in Fig. 3) is fabricated from nickel foil. The air-side flow field is ferritic stainless steel. The interconnect plates and flow fields also serve as electrical conductors and current distributors. To improve performance, the air-side separator plates and flow fields are pre-surface-treated to form a rare-earth conductive oxide scale. A perovskite rare-earth coating is also applied to the separator-plate oxide scale by either screen printing or plasma spraying. On the steam/hydrogen side of the separator plate, a thin (~10 μm) nickel metal coating is applied.

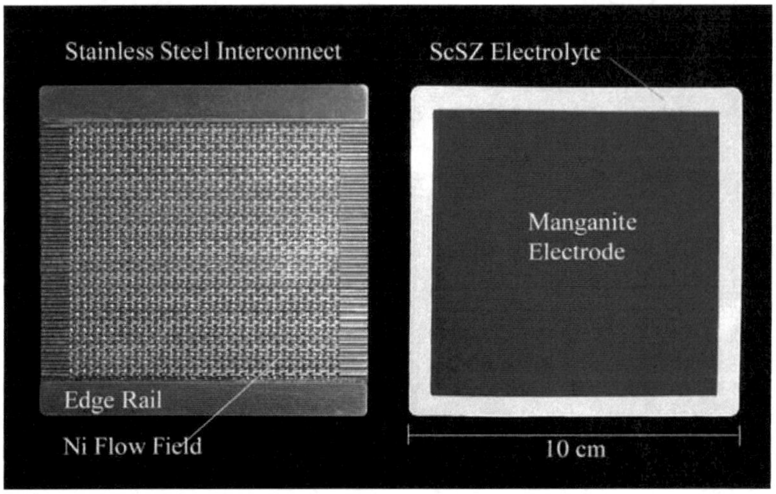

Figure 3. Interconnect plate and single electrolysis cell.

The electrolyte / electrode assembly is shown on the right of Fig. 3. The electrolyte is scandia-stabilized zirconia, about 140 μm thick. The air-side electrode (anode in the electrolysis mode), visible in the figure, is a strontium-doped manganite. The electrode is graded, with an inner layer of manganite/ zirconia (~13 μm) immediately adjacent to the electrolyte, a middle layer of pure manganite (~18 μm), and an outer cobaltite bond layer. The steam/ hydrogen electrode (cathode in the electrolysis mode) is also graded, with a nickel-zirconia cermet layer (~13 μm) immediately adjacent to the electrolyte and a pure nickel outer layer (~10 μm).

Measurement of the outlet dewpoint temperature, downstream of the electrolysis stack, allows for direct determination of the change in dewpoint, the rate of steam reduction and the corresponding rate of hydrogen production during electrolysis testing. All testing was performed at essentially ambient laboratory pressure, about 85 kPa for our 4700 ft. elevation.

Stack voltage and current values were controlled by means of a programmable DC power supply (Lambda Zup 20-40), rated for 20 V and 40 amps. The power supply could be operated in either the constant-current or constant voltage modes. This power supply was interfaced to the system controller computer via an RS232 interface. Cell amperage and the voltage across the stack are monitored continuously. Stack current measurements are obtained from the power supply, with redundant measurements provided by a precision shunt (50 A = 200 mV).

Signals from all thermocouple channels, mass-flow controllers, pressure transducer, and stack voltage taps were wired into a modular data acquisition system (National Instruments Model SCXI) which was configured for this application with a 32-channel analog input module and an eight-channel analog output module. The analog output module was used to provide control signals (4 – 20 mA) for feedback temperature control of the humidifier, the humidity measurement stations, and the furnace. Data acquisition and instrument control were implemented for these tests using a Labview (National Instruments, version 6.1) program (virtual instrument) specifically created for this experiment. Temperature control of the humidifier, the humidity sensing volumes, and the furnace was accomplished using customized PID control algorithms.

DATA REDUCTION

Open-cell potentials can be predicted for any gas composition from the Nernst equation, which for the hydrogen / oxygen / steam system takes the form:

$$E = \frac{-\Delta \bar{g}_f^o}{2F} - \frac{R_u T}{2F} \ln\left[\left(\frac{y_{H2O}}{y_{H2} y_{O2}^{1/2}}\right)\left(\frac{P}{P_{std}}\right)^{-1/2}\right] \tag{1}$$

Since we are using air as our source of oxygen, the value of y_{O2} at the stack inlet is fixed at 0.21. In the stack configuration, potentials are usually measured across the entire stack, so the overall stack open-cell potential is given by:

$$E_{stack} = N_{cells} E \tag{2}$$

where E is the single-cell open-cell potential, from Eqn. (1).

Inlet and outlet gas stream dewpoint temperatures were directly measured in this work. Aside from the fact that the measured difference between outlet and inlet dewpoint temperature (dewpoint depression) is a useful indication of cell performance, measured dewpoints permit direct determination of inlet and outlet steam mole fractions and the corresponding hydrogen production rate. From the measured dewpoint temperatures, water vapor pressures may be calculated from an appropriate correlation (e.g., Antoine). The inlet and outlet mole fractions of steam can then be obtained from:

$$y_{H2O,i} = \frac{P_{H2O,i}}{P}; \qquad y_{H2O,o} = \frac{P_{H2O,o}}{P}. \tag{3}$$

It should be noted that since the mole fraction of steam in the inlet gas flow is determined strictly by the dewpoint temperature, the total inlet flow rate of steam is therefore directly proportional to the sum of the molar flow rates of nitrogen plus hydrogen:

$$\dot{N}_{H2O,i} = \frac{\left(\dot{N}_{N2} + \dot{N}_{H2,i}\right) y_{H2O,i}}{1 - y_{H2O,i}}. \tag{4}$$

The magnitude of the total inlet steam flow rate is important in determining whether or not steam starvation is likely to occur during electrolysis operation.

The inlet mole fraction of hydrogen, $y_{H2,i}$ is determined by the steam mole fraction and the ratio of the nitrogen flow rate to the hydrogen flow rate, both of which are determined by the setpoint values established on the mass flow controllers:

$$y_{H2,i} = \frac{1 - y_{H2O,i}}{1 + \dot{N}_{N2} / \dot{N}_{H2,i}} \tag{5}$$

$$\dot{N}_{N2} = Q_{s,N2} \frac{P_{std}}{R_u T_{std}} \tag{6}$$

$$\dot{N}_{H2,i} = Q_{s,H2} \frac{P_{std}}{R_u T_{std}}. \quad (7)$$

Finally, the molar rates of hydrogen production and steam consumption may be determined from:

$$\Delta\dot{N}_{H2} = -\Delta\dot{N}_{H2O} \quad (8)$$

$$\Delta\dot{N}_{H2} = \dot{N}_{H2O,i} - \left(\dot{N}_{H2O,i} + \dot{N}_{H2,i} + \dot{N}_{N2}\right) y_{H2O,o}. \quad (9)$$

The molar rates of hydrogen production and steam consumption during electrolysis can also be predicted independently from the measured stack electrical current:

$$\Delta\dot{N}_{H2} = \frac{I}{2F} N_{cells} = -\Delta\dot{N}_{H2O} \quad (10)$$

where N_{cells} is the number of cells in the stack. The product of the electrical current and the number of cells is the total ionic current in the stack.

Since we have directly measured both inlet and outlet dewpoint temperatures, we have generated plots comparing hydrogen production rates based on stack current to hydrogen production rates based on the measured change in dewpoint over a range of stack current densities.

An important performance parameter that quantifies the ohmic losses associated with the operation of solid-oxide electrolysis cells is the area-specific resistance (ASR). This quantity is defined as:

$$ASR = \frac{E - E_{OCV}}{i}; \quad i = \frac{I}{A_{cell}} \quad (11)$$

where E is the applied potential, E_{OCV} is the Nernst or open-cell potential and i is the current density (A/cm^2).

In the electrolysis mode of operation, the electrochemical reaction is endothermic, and ohmic heating is always present. At operating voltages between the open-cell potential and the thermal neutral voltage, the endothermic reaction heat requirement is greater in magnitude than the ohmic heating, and a net cooling effect prevails in the stack (O'Brien et al., 2004). The thermal neutral voltage, given by:

$$V_{tn} = \frac{\Delta h_R}{2F} \quad (12)$$

is 1.287 V at 800ºC, or 12.87 V for the 10-cell stack. Beyond this operating voltage, ohmic heating exceeds the endothermic reaction heat requirement and a net heating effect prevails.

NUMERICAL MODEL

The numerical model developed for this paper was based on the geometry of a single SOEC cell taken from the stack described previously. The numerical domain extends from the center plane of one separator plate to the center plane of the next separator plate. Symmetry boundaries are applied at the top and bottom of the model. Representations of the numerical model are presented in Figure 4. In the top left portion of Figure 4, the full model is shown to scale. Since the model includes only one cell, the model geometry is quite thin in the vertical (z) direction. To show more detail, the model is shown in the bottom left portion of Figure 4 with a vertical exaggeration of 10x in the z-direction. An exploded view with the 10x vertical exaggeration is shown in the right half of the figure.

In the exploded view, the bottom element is the bottom separator plate. Since we are trying to represent a unit cell extracted from a larger stack, the bottom and top separator plates in the numerical model are only half as thick (i.e., 0.19 mm) as the hardware separator plates. Therefore, the top and bottom boundaries of the numerical model represent symmetry planes and the boundary conditions on those faces are set accordingly. The edge rails are shown attached to the bottom separator plate. In the stack hardware, the edge rails are fabricated from the same material as the separator plates, but they are separate pieces.

The next element in the numerical model is the steam/hydrogen flow channel. The flow channels are the regions in the stack between the separator plate, the edge rails and the electrodes in which the corrugated/perforated "flow fields" are located. The steam/hydrogen flow channel has been specified as a high-porosity porous-media region with metallic nickel as the solid material and with anisotropic permeability, much higher in the primary flow direction than in the cross flow directions. The height of the flow channel is the set by the thickness of the edge rails, 1.019 mm.

The next three layers in the numerical model are associated with the electrolyte/electrode assembly, as shown in the right half of Figure 4. The FLUENT SOFC module treats the electrolyte as a 2-D planar element. Therefore the electrolyte in the model has geometrical thickness of zero. On either side of the electrolyte are the electrodes which are created with 3-D elements. Therefore, the electrolyte/electrode assembly in the model is only as thick as the two electrodes. Around the outer periphery of the electrolyte/electrode assembly, we have include an "insulator" with the properties of YSZ. The insulator prevents an electrical short circuit between the top and bottom edge rails. No ionic transport occurs through this insulator.

The next element in the numerical model is the air/oxygen flow channel. It has also been specified as a high-porosity porous media region with ferritic stainless steel as the solid material and with the same anisotropic permeabilities and flow channel height used in the steam/hydrogen flow channel. The top separator plate and edge rails are identical to those on the bottom, but the edge rails are oriented perpendicular to the bottom edge rails to allow for the cross flow arrangement. The bottom separator plate in the FLUENT model serves as the electrical ground and the top separator plate serves as the current source.

Additional parameters specified in the numerical model include the electrode exchange current densities and several gap electrical contact resistances. These quantities were determined empirically by comparing FLUENT predictions with stack performance data. The FLUENT model uses the electrode exchange current densities to quantify the magnitude of the activation overpotentials via a Butler-Volmer equation (Prinkey et al., 2004). A radiation heat transfer boundary condition was applied around the periphery of the model to simulate the thermal conditions of our experimental stack, situated in a high-temperature electrically heated radiant furnace. The edges of the numerical model are treated as a small surface in a large

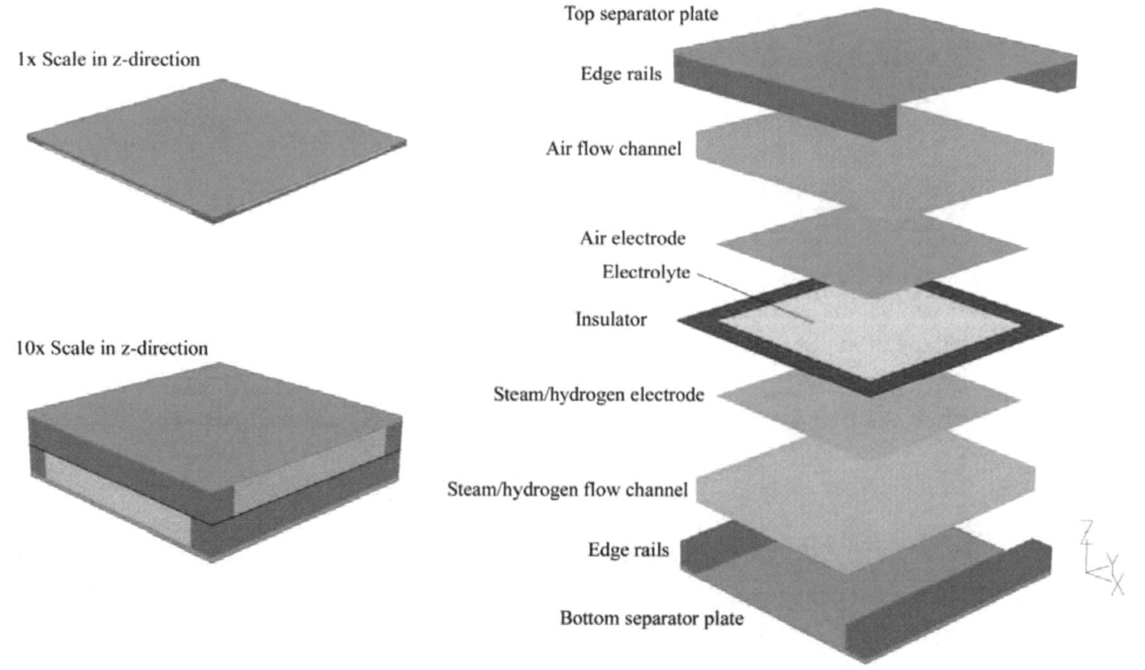

Figure 4. Fluent single-cell SOEC model.

enclosure with an effective emissivity of 1.0, subjected to a radiant temperature of 1103 K, equal to the gas-inlet temperatures.

The gas flow inlets are specified in the FLUENT model as mass-flow inlets, with the gas inlet temperatures are set at 1103 K and the inlet gas composition determined by specification of the mass fraction of each component. The gas flow rates used in the model were the same as those used for the experimental base case, on a per-cell basis. For example, the base case for the steam/hydrogen inlet used a total inlet mass flow rate of 8.053 x 10^{-6} kg/s, with nitrogen, hydrogen and steam mass fractions of 0.51, 0.0074, and 0.483, respectively. The base case air flow rate was 4.33 x 10^{-6} kg/s.

Details of the core mass, momentum, energy, and species conservation and transport features of FLUENT are documented in detail in the FLUENT user manual (FLUENT, 2004). Details of the electrochemical reactions, loss mechanisms, electric field computation, and electrode porous media constitutive relations are documented by Prinkey et al. (2004). This reference also documents the treatment of species and energy sources and sinks arising from the electrochemistry at the electrode-electrolyte interfaces.

RESULTS

A series of DC potential sweeps was acquired in order to characterize the performance of the stack over a range of operating conditions. Each sweep was performed in the electrolysis mode with the furnace temperature and gas flow rates set at fixed values, but with the power supply programmed to vary the applied stack voltage over a range, typically from 8 to 14 volts. The power supply sweep rate was set at ~8 mV/s, so the duration of each sweep was about 25 minutes. At the gas flow rates used in the stack, the response time of the downstream dewpoint sensor was fast enough to provide continuous outlet dewpoint measurements during

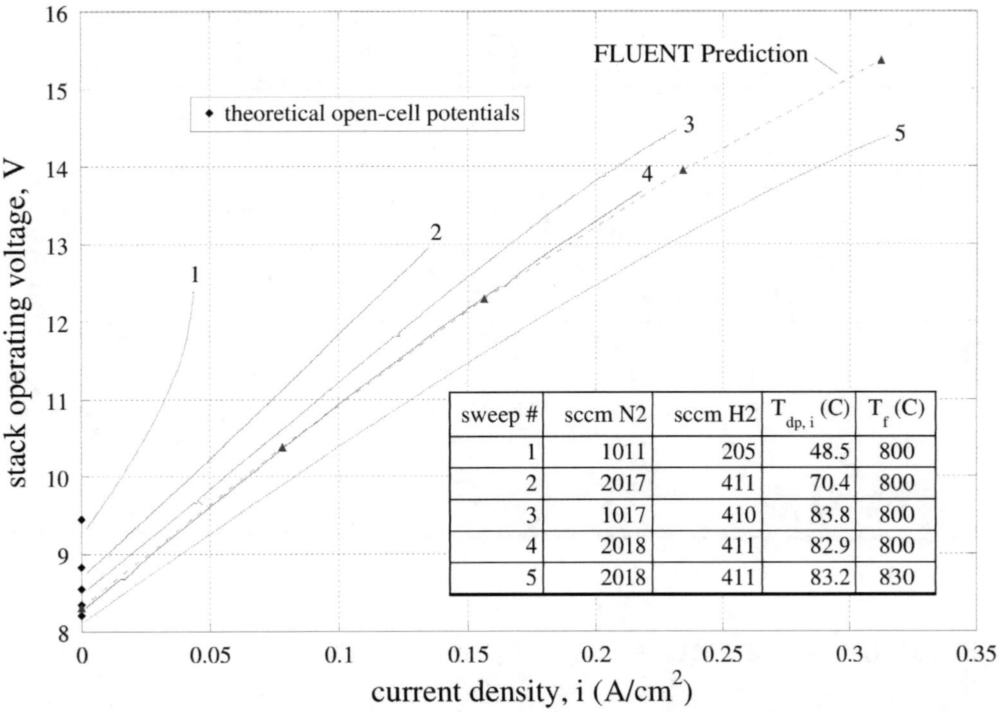

sweep #	sccm N2	sccm H2	$T_{dp,i}$ (C)	T_f (C)
1	1011	205	48.5	800
2	2017	411	70.4	800
3	1017	410	83.8	800
4	2018	411	82.9	800
5	2018	411	83.2	830

Figure 5. Stack operating potential as a function of current density.

the sweep. These sweeps were useful for obtaining information about cell area-specific resistance (ASR) values and steam starvation limits.

Results of several representative sweeps are shown in Fig. 5 in the form of operating voltage versus current density. Test conditions for each of the five experimental sweeps shown are tabulated in the figure. Four of the sweeps were obtained at a furnace temperature of 800ºC and one at 830ºC. Inlet dewpoint values were varied as shown. Theoretical open-cell potential values are also shown in the figure for each sweep using a single data point at zero current density. Note that the measured open-cell potentials are in excellent agreement with these theoretical values for each sweep. In additional to the experimental curves, a FLUENT prediction is also shown. This prediction was obtained for the conditions of sweep 4, with the gap contact resistance values empirically adjusted to yield an overall area-specific resistance value that matches the experimentally observed value. Additional FLUENT predictions of local temperatures and current densities presented later were obtained using the same procedure.

Sweep #1 was performed with a relatively low inlet steam flow rate, corresponding to the low inlet dewpoint value of 48.5ºC and relatively low nitrogen and hydrogen flow rates. This sweep has a relatively high slope on i-V coordinates, indicating a high ASR value. This sweep also clearly shows the effects of steam starvation; the slope of the i-V curve increases dramatically as the current density is increased. The outlet dewpoint temperature corresponding to the highest current density shown in this figure was only 4ºC. Sweep #2 was performed at an intermediate steam concentration, with an inlet dewpoint temperature of 70ºC. This sweep exhibits nearly linear behavior over the range of current densities shown, with a much smaller slope than sweep #1. Sweeps 3 and 4 are nearly linear at low current densities, then slightly concave-down at higher current densities. Sweep #5 has a shallower slope than

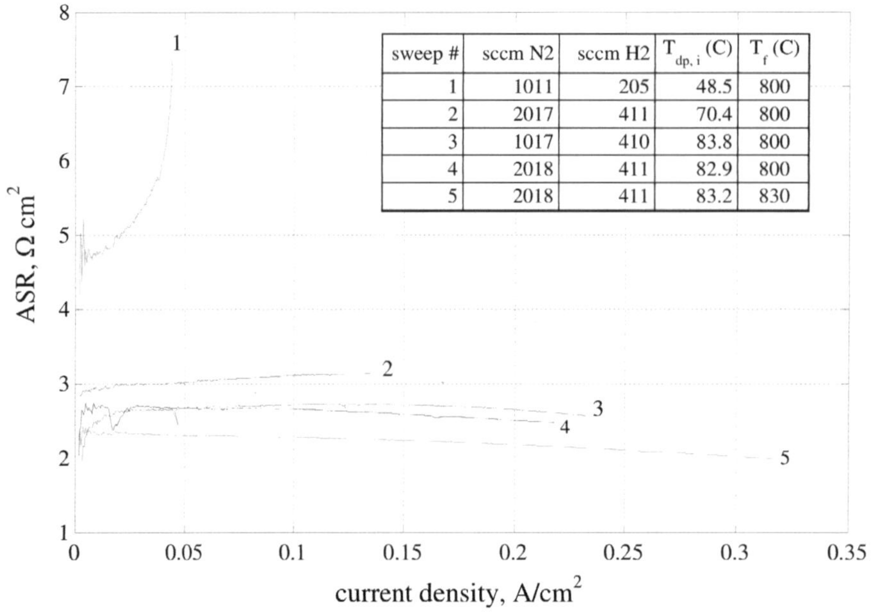

Figure 6. Per-cell area-specific resistance measured during DC potential

the others, consistent with the higher operating temperature of 830ºC. This i-V curve also shows more distinct concave-down characteristics at higher current densities.

Per-cell area-specific resistance values corresponding to the sweeps of Fig. 5 are presented in Fig. 6 as a function of current density. Highest ASR values were observed for the cases with low steam concentrations. Sweep #1 had ASR values near 4.5 at low current density, increasing rapidly for higher current densities, due to steam starvation. Sweep #2, at intermediate steam concentration had an intermediate ASR value near 3.0, increasing slightly with increasing current density. Sweeps 3 and 4 were similar, with ASR values near 2.5, decreasing slightly at high current density. Sweep #5, at an operating temperature of 830ºC showed the lowest ASR value, around 2.2, decreasing significantly with increasing current density.

Hydrogen production rates can be calculated directly from the stack electric current (Eqn. 10) and independently from the measured inlet and outlet dewpoint measurements (Eqn. 9). A representative plot of hydrogen production during DC potential sweep #2 is shown in Fig. 7. The left-hand vertical scale is in sccm and the right-hand vertical axis is in NL/hr. The current-based hydrogen production rate is simply a straight line since hydrogen production is directly proportional to the current. The dewpoint-based measurement shows some scatter associated with the instantaneous measured inlet and outlet dewpoint values. Agreement between the two measurements is generally very good. Hydrogen production rates as high as 90 NL/hr were achieved with this stack.

Temperatures measured during DC potential sweep #4 are presented in Fig. 8 as a function of operating voltage. Stack internal temperature measurements were obtained using miniature thermocouples inserted into the air flow channels. Stack internal #1 was located in the center position of the center air flow channel, stack internal #2 was located along the edge of the center cell closest to the steam/hydrogen exit, stack internal #3 was located along the edge of the center cell closest to the steam/hydrogen inlet, and stack internal #4 was located

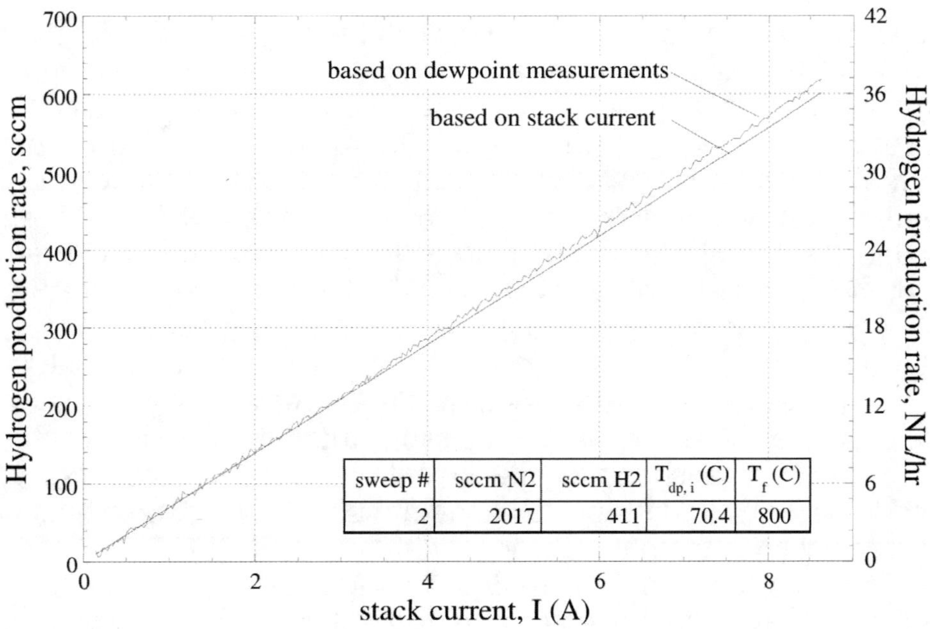

Figure 7. Hydrogen production rates during DC potential sweep.

at the center of the top air flow channel. The air inlet temperature is obtained from a thermocouple located inside the air inlet manifold. This thermocouple is used as the process variable for feedback control of the furnace power. The "furnace" temperature is obtained form a thermocouple located in the main part of the furnace outside of the test fixture.

The air inlet temperature remains virtually fixed during the DC potential sweep because this temperature is used for feedback control to maintain a constant stack operating temperature. The four internal stack thermocouples respond as expected during the sweep.

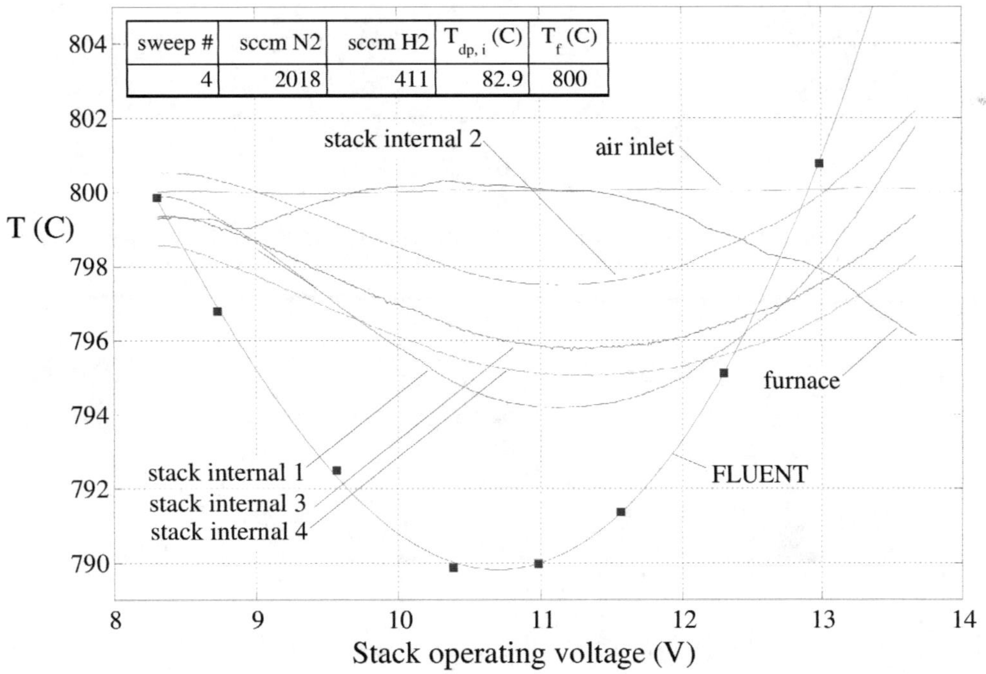

Figure 8. Stack temperatures during DC potential sweep.

At voltages between the open-cell potential and the thermal neutral voltage, in the electrolysis mode, the stack internal temperatures are lower than the gas inlet and furnace setpoint temperatures because at this operating condition, the endothermic reaction heat requirement is greater than the ohmic heating (O'Brien et al., 2004) and there is a net cooling effect on the stack. A thermal minimum point is reached at an operating voltage near 11.2 V and full thermal recovery is observed near the thermal neutral voltage of 12.9 V. The magnitude of the stack cooling and heating effects is greatest at the center of the stack, as indicated by the response of internal stack temperature 1. Internal stack temperatures 2 and 3 respond similarly to the center thermocouple, but with smaller amplitude, due to convective and radiative heat transfer effects near the edges of the stack. Stack internal #4 is in the top air flow channel, so its response is also constrained by radiant heat transfer. Note that the "furnace" thermocouple responds in a direction that is opposite of the stack internal thermocouples. This is because the feedback control demands more furnace power when the stack is tending to cool off and less power when the stack is tending to heat up. Figure 8 also shows the mean electrolyte temperature predicted from a FLUENT simulation obtained at the operating conditions of sweep #4. The trend of the mean electrolyte temperatures with respect to operating voltage is similar to the trend observed with the thermocouple readings. However, the magnitude of the predicted temperature variation is greater than what was measured. This could due to the fact that the FLUENT model, as it was configured for these calculations, does not include internal radiation heat transfer. We plan to investigate this effect further in our future modeling efforts.

Contour plots representing local FLUENT results for electrolyte temperature and current density are presented in Figs. 9 and 10. In these figures, the steam/hydrogen flow is from top to bottom and the air flow is from left to right. Figure 9 shows electrolyte temperature contour plots for amperages of 10, 15, and 30 amps. These current values correspond to

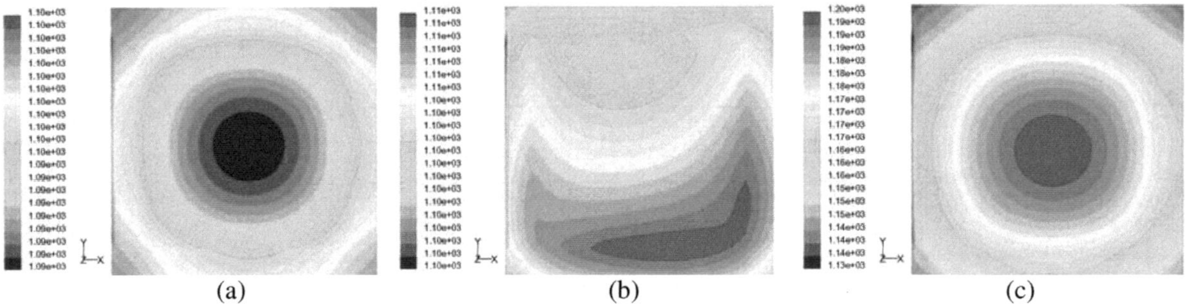

(a) (b) (c)

Figure 9. Temperature (K) contours on the electrolyte and insulator for currents of 10, 15, and 30 amps.

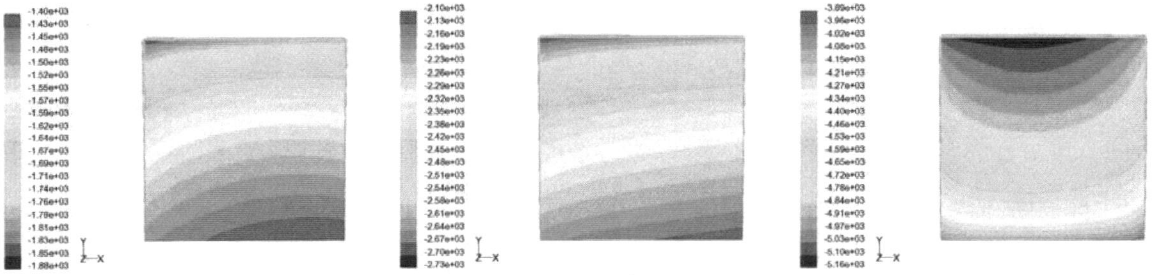

(a) (b) (c)

Figure 10. Current density (A/m^2) contours on the electrolyte for currents of 10, 15, and 30 amps.

operating voltage regions shown on Figure 8 near the minimum electrolyte temperature (10 amps), near thermal neutral voltage (15 amps), and in the region dominated by ohmic heating (30 amps). The radiant boundary condition at 1103 K tends to hold the outside of the model at a higher temperature for the 10-amp case (Fig. 9 (a)), while the endothermic heat requirement maintains the center of the electrolyte at a lower temperature. Minimum and maximum temperatures for this case are 1091 K and 1100 K respectively. The center Fig. 9 (b) shows a temperature difference across the electrolyte of only one degree K, with values very near 1103 K; this current density is very near the thermal neutral voltage. Fig. 9 (c) shows that ohmic heating in the electrolyte is dominating and the thermal boundary condition is keeping the edges cooler than the inside. Minimum and maximum temperatures are 1139 K and 1197 K, respectively, for this case.

Contour plots of local current density on the electrolyte are shown in Figure 10 for 10, 15, and 30 amps. Mean current densities for these three cases are: 0.156, 0.234, and 0.469 A/cm2. These plots correlate directly with local hydrogen production rates. Since FLUENT is being run in electrolysis mode, the current density values are all negative and hence the blue values have the largest magnitudes. Highest current density magnitudes occur near the steam hydrogen inlet (the top of the figures). This corresponds to the location of the greatest steam concentration. The orange areas show where the current density is lowest because the available steam concentration is lower.

SUMMARY AND CONCLUSIONS

A research program is under way at the Idaho National Laboratory to assess the performance of solid-oxide cells operating in the steam electrolysis mode for hydrogen production over a temperature range of 800 to 900ºC. The research program includes both experimental and modeling activities. Selected results from both activities have been presented in this paper. Experimental results were obtained from a ten-cell planar electrolysis stack, with a per-cell active area of 64 cm^2, fabricated by Ceramatec, Inc. The electrolysis cells are electrolyte-supported, with scandia-stabilized zirconia electrolytes (~140 µm thick), nickel-cermet steam/hydrogen electrodes, and manganite air-side electrodes. The metallic interconnect plates are fabricated from ferritic stainless steel. The experiments were performed over a range of steam inlet mole fractions (0.1 – 0.6), gas flow rates (1000 – 4000 sccm), and current densities (0 to 0.38 A/cm^2). Steam consumption rates associated with electrolysis were measured directly using inlet and outlet dewpoint instrumentation. Cell operating potentials and cell current were varied using a programmable power supply. DC potential sweeps performed on this stack demonstrated the effects of steam flow rate on stack performance. High ASR values and steam starvation were observed for the low-steam-content sweeps. A per-cell ASR value of 2.2 was demonstrated for a high-steam-content DC potential sweep at 830ºC. Hydrogen production rates measured by the change in dewpoint of the gas flows were shown to be in excellent agreement with hydrogen production rates based on stack current. Hydrogen production rates up to 90 Normal liters per hour were demonstrated. Internal stack temperatures measured with miniature thermocouples A stack cooling effect was directly observed for stack operating voltages between the open-cell potential and the thermal neutral voltage.

A three-dimensional computational fluid dynamics (CFD) model was also created to model high-temperature steam electrolysis in a planar solid oxide electrolysis cell (SOEC). The model represents a single cell as it would exist in the experimental electrolysis stack.

Mass, momentum, energy, and species conservation and transport are provided via the core features of the commercial CFD code FLUENT. A solid-oxide fuel cell (SOFC) model adds the electrochemical reactions and loss mechanisms and computation of the electric field throughout the cell. The FLUENT SOFC user-defined subroutine was modified for this work to allow for operation in the SOEC mode. Model results provide detailed profiles of temperature, Nernst potential, operating potential, anode-side gas composition, cathode-side gas composition, current density and hydrogen production over a range of stack operating conditions. Mean model results were shown to compare favorably with the experimental results obtained from the ten-cell stack tested at INL. At operating voltages between the open-cell potential and thermal neutral, electrolyte temperatures are below 1103 K. A thermal minimum temperature occurs near 1.08 V. For operating voltages above thermal neutral, ohmic heating dominates and resultant electrolyte temperatures rapidly increase beyond 1103 K. Contour plots of local electrolyte temperature and current density indicated the effects of heat transfer, reaction cooling/heating, and change in local gas composition.

ACKNOWLEDGEMENTS

This work was supported by the US Department of Energy, Office of Nuclear Energy, Nuclear Hydrogen Initiative Program.

REFERENCES

FLUENT Theory Manual, version 6.1.22, Fluent Inc., Lebanon, New Hampshire, 2004.

Herring, J. S., O'Brien, J. E., Stoots, C. M., Lessing, P. A., Anderson, R. P., Hartvigsen, J. J., and Elangovan, S., "Hydrogen Production from Nuclear Energy via High-Temperature Electrolysis," presented at the 2004 International Conference on Advances in Nuclear Power Plants (ICAPP '04), June 13-17, 2004, Pittsburgh, PA.

Herring, J. S., O'Brien, J. E., Stoots, C. M., Lessing, P. A., Anderson, R. P., Hartvigsen, J. J., and Elangovan, S., "Hydrogen Production through High-Temperature Electrolysis Using Nuclear Power," presented at the AIChE Spring National Meeting, New Orleans, April 25 – 29, 2004.

International Atomic Energy Agency (IAEA), May 1999, *Hydrogen as an energy carrier and its production by nuclear power*, IAEA-TECDOC-1085.

National Academy of Sciences, National Research Council, *The Hydrogen Economy: Opportunities, Costs, Barriers, and R&D Needs*, February, 2004.

O'Brien, J. E., Stoots, C. M., Herring, J. S., and Lessing, P. A., "Characterization of Solid-Oxide Electrolysis Cells for Hydrogen Production via High-Temperature Steam Electrolysis," Proceedings, 2nd International Conference on Fuel Cell Science, Engineering, and Technology, June 14-16, 2004, Rochester, NY, paper# 2474, pp., 219 – 228.

O'Brien, J. E., Stoots, C. M., Herring, J. S., and Lessing, P. A., "Performance Measurements of Solid-Oxide Electrolysis Cells for Hydrogen Production from Nuclear Energy," Proceedings, 12th ICONE Meeting, April 25-29, 2004, Arlington, VA, paper # ICONE12-49479.

Prinkey, M., Shahnam, M., and Rogers, W. A., "SOFC FLUENT Model Theory Guide and User Manual," Release Version 1.0, FLUENT, Inc., 2004.

Yildiz, B., and Kazimi, M. S., "Nuclear Energy Options for Hydrogen and Hydrogen-Based Liquid Fuels Production," MIT-NES-TR-001, September 2003.

THERMAL-FLUID AND ELECTROCHEMICAL MODELING AND PERFORMANCE STUDY OF A PLANAR SOLID OXIDE ELECTROLYSIS CELL

Bilge Yildiz, Tanju Sofu
Argonne National Laboratory, Nuclear Engineering Division,
9700 S Cass Avenue, Bldg 208, Argonne, IL 60439
byildiz@anl.gov, tsofu@anl.gov

INTRODUCTION

The goal of this project is to develop a model that can guide the performance optimization of high temperature electrolysis (HTE) systems for hydrogen production. The currently considered HTE system concepts are based on solid oxide electrolysis cells (SOECs) that are similar to those used for solid oxide fuel cells (SOFCs). The SOECs operate with an applied potential to electrocatalytically split steam into $H_{2(g)}$ and $O_{2(g)}$. There is limited knowledge about the intermediate reaction steps that define the phenomenological behavior of solid oxide electro-chemistry, especially for the less investigated SOEC mode of operation. Several electrochemical and thermal-fluid dynamics models have been developed for SOFCs and some of them can be found in the Refs. 1-3. However, none has been built for SOECs based on the first-principals of the governing electrochemical mechanisms. The model developed in this work appropriately couples the electrochemical mechanisms to the heat transfer and fluid dynamics in the operation of SOECs to guide the design and optimization of SOEC based hydrogen production systems.

THERMAL-FLUID AND ELECTROCHEMICAL MODEL

The model combines an in-house developed electrochemical (EC) module and the commercially available computational fluid dynamics (CFD) code STAR-CD. The electrochemical model includes an assessment of the minimum required electrical potential (Nernst potential), V_{min}, as well as the ohmic, η_Ω, activation, η_{act}, and diffusion, η_{diff}, related overpotentials in the cell components as a function of the current density, i, such that the relation in Eq 1 is satisfied.

$$V_{appl} - V_{min} = \eta_{act}^{anode}(i) + \eta_{act}^{cathode}(i) + \eta_{diff}^{anode}(i) + \eta_{diff}^{cathode}(i) + \eta_{\Omega}^{anode}(i) + \eta_{\Omega}^{cathode}(i) + \eta_{\Omega}^{electrolyte}(i) \quad [1]$$

This leads to evaluation of the current density distribution, heat sources and sinks, and $H_2O_{(g)}$, $H_{2(g)}$ and $O_{2(g)}$ sources and sinks within the SOEC. The CFD module performs thermal and flow analysis based on the specified boundary conditions, and the heat and species generation rates calculated by the EC module. The flow in the porous electrode and flow-mesh regions of the SOEC is modeled as a compressible multi-component mixture with relevant heat and mass sources and sinks. The main coupling between the EC and CFD modules is due to the mutual use of the temperature profile. The CFD module provides the temperature field for the EC module to generate the current density distribution through using the temperature dependent electrochemical parameters. The EC module provides the heat generation rate, based on the current density that it calculates, for the CFD module to generate the consistent temperature profile.

SIMULATIONS

A planar geometry SOEC with cross flow configuration is considered in this study as the initial application. The geometric parameters, flow field boundary conditions and configuration, and material properties of the electrolyte-supported cell-stack currently being tested at Idaho National Laboratory (INL) make the input to our initial simulations. The electrochemically active area of the cells is 8x8 cm^2. The schematic for the cross-flow configuration of the SOEC is shown in Fig. 1. The electrolyte, the oxygen electrode (anode), and hydrogen electrode (cathode) are made of ytrria-stabilized zirconia (YSZ), $La_{(1-x)}Sr_{(x)}MnO_3$ (LSM), and Ni-YSZ, respectively. The hydrogen electrode has a mixture of steam and hydrogen of $4x10^{-6}$ kg/s, 800°C and atmospheric pressure at its inlet. The mass fraction of hydrogen at the cathode inlet is 1.22% at INL experiments, and 0% and 1.22% at our initial simulations. The oxygen electrode has air of $3.28x10^{-6}$ kg/s, 800°C and atmospheric pressure at its inlet. Simulations were performed with the above stated operating conditions for a range of applied cell-potential values in order to determine the cell efficiency, hydrogen production rate, and temperature and current distribution profiles.

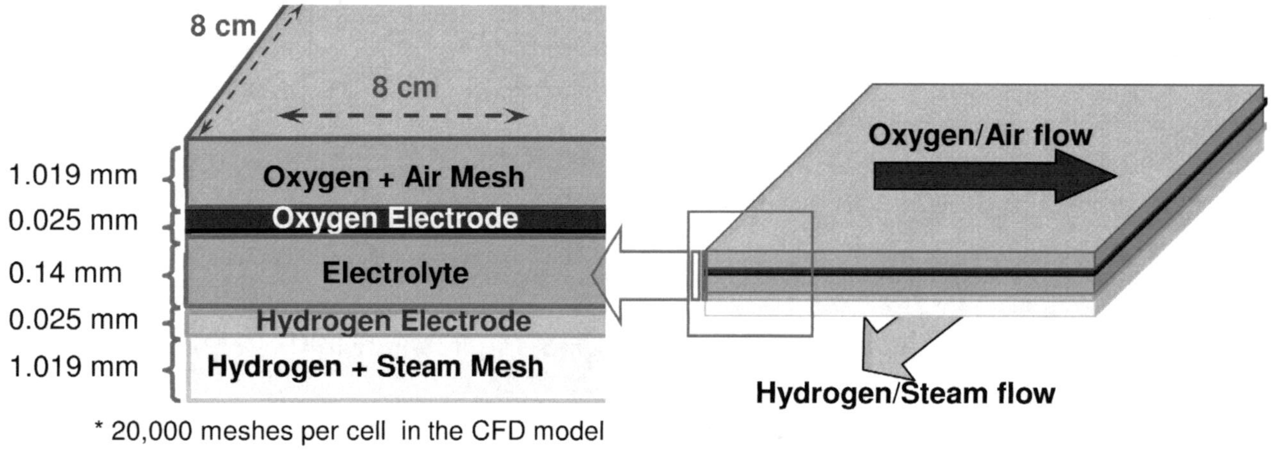

Figure 1: The schematic for the cross-flow configuration of SOEC used in the simulations (not-to-scale).

RESULTS

The polarization behavior of the simulated cell is studied within the applied potential range of 0.8 – 1.4 V. For the operating conditions used for this model, the given range of applied potential can yield 0 – 5000 A/m^2 of average current density and a corresponding hydrogen production density, with an assumed limiting current density of 6000 A/m^2. The typical temperature, current density, H_2 concentration, and minimum required potential profiles are shown in Fig. 2 for an applied cell-potential of 1.15 V on the cross-flow SOEC. The 1.15V of applied potential is slightly above the resulting thermal neutral point of the cell. The resulting polarization, Nernst potential, voltage efficiency and cell average temperature curves for the modeled cell, with 0% and 1.22% of hydrogen mass fraction at the cathode inlet, are presented in Fig. 3. This figure indicates that the H_2 fraction at the inlet can influence the efficiency and the cost of the SOEC system. A higher hydrogen fraction at the inlet indicates higher voltage efficiency for the cell (Fig.3-c). Nevertheless, at a given current density (hydrogen production density), operating with no hydrogen at the inlet would require less of an applied potential on the cell (Fig.3-a) and improve the overall efficiency of the system. In spite of this, an amount of hydrogen at the inlet can be required at actual operating conditions to avoid the oxidation and long-term degradation of the Ni catalyst in the hydrogen electrode.

The results indicate that the mass transfer limited region is close to 1.35 V with an assumed limiting current of 6000 A/m^2. The numerical convergence of the program for applied potential values of above 1.35 V has been unfeasible as of yet. This is due to very steep gradient of the voltage-current density relationship as the current reaches the mass transfer limited region. The region where the ohmic processes govern the polarization behavior corresponds approximately to 1.0 – 1.3 V. The resulting current density in this operating regime varies from 800 to 4100 A/m^2, corresponding to a hydrogen production rate of 0.716 – 3.670 kgH_2/m^2/day. The voltage efficiency drops from 87% at 1.0 V down to 69% at 1.3, for 1.22% hydrogen at the cathode inlet.

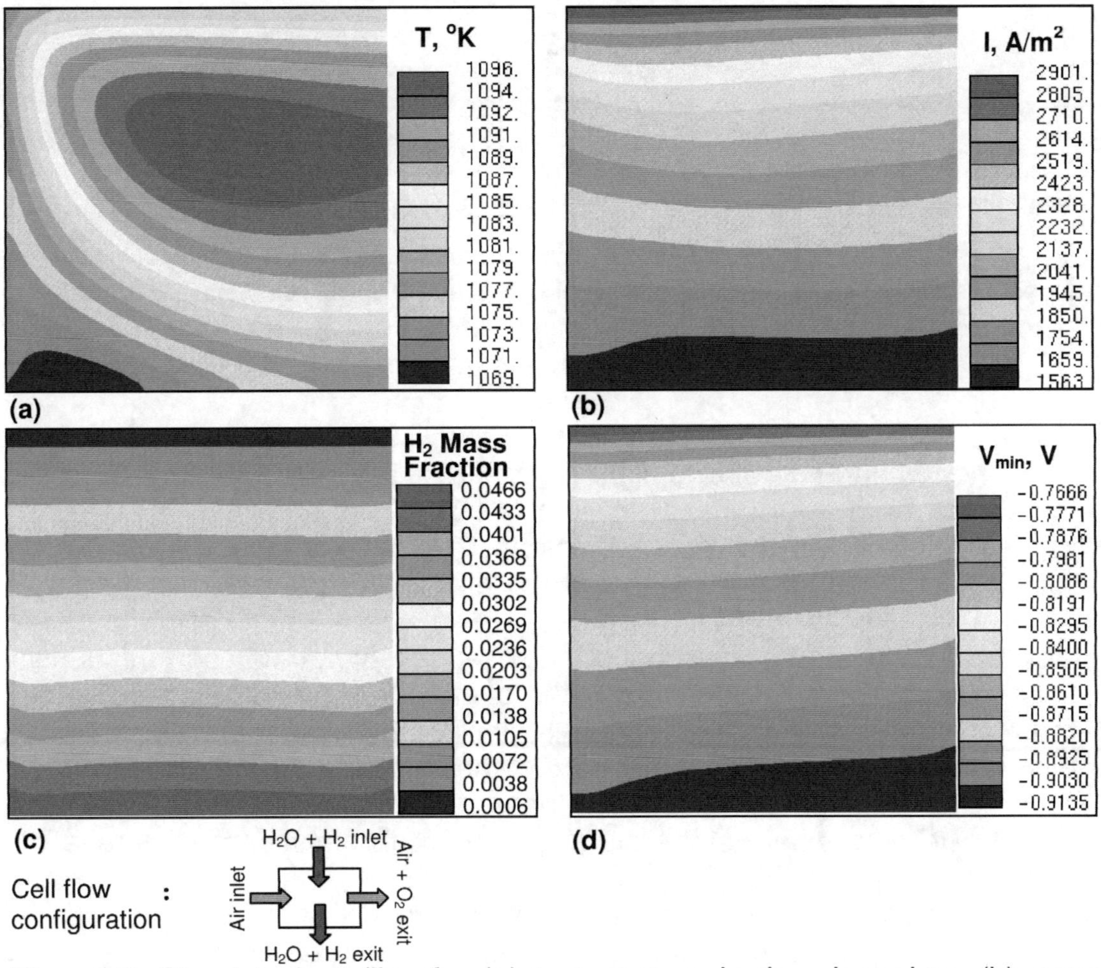

Figure 2. Simulated profiles for (a) temperature in the electrolyte, (b) current density, (c) hydrogen mass fraction in the cathode mesh, (d) minimum required potential (Nernst potential) for the cross-flow configuration SOEC. Inlet conditions: 800°C and 1atm, $V_{applied}$: 1.15V, with 0% of H_2 at the cathode inlet.

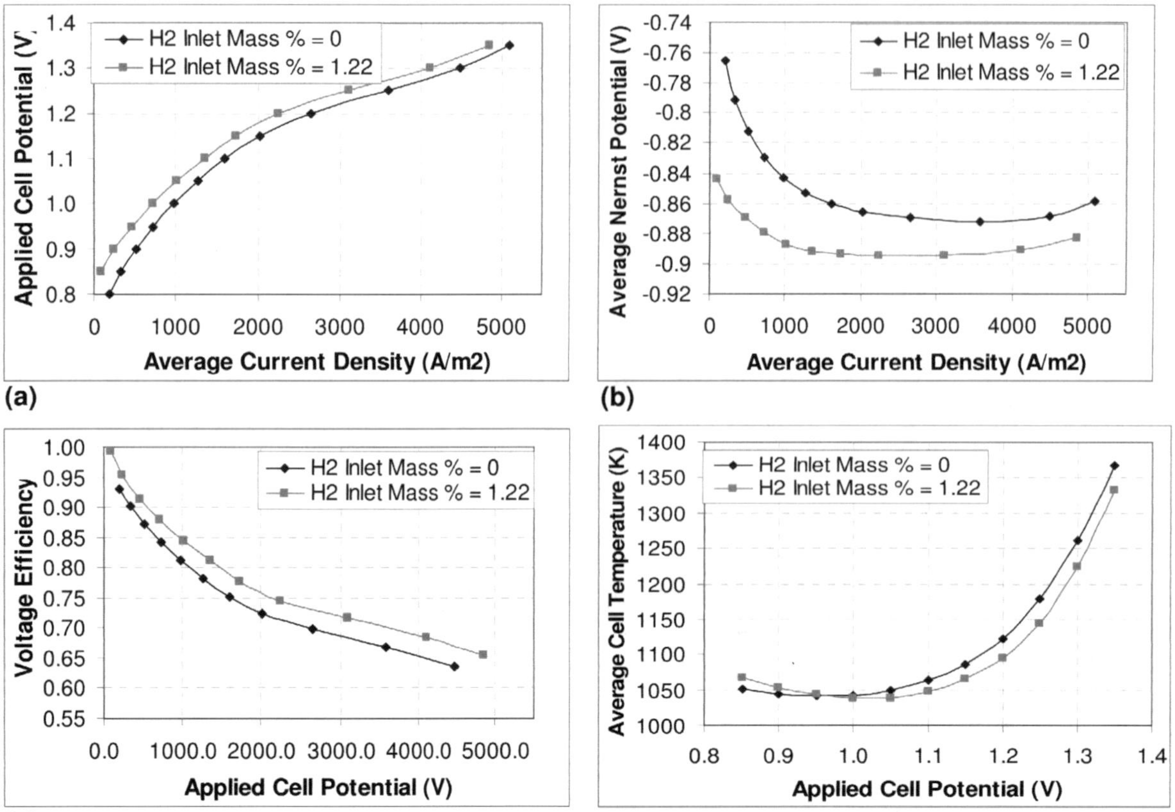

(c) (d)

Figure 3. Simulated results of (a) polarization, (b) Nernst Potential, (c) voltage efficiency, and (d) average cell temperature for the cross-flow configuration SOEC. Inlet conditions: 800°C and 1atm, $V_{applied}$:1.15V, with 0% and 1.22% mass fraction of H_2 at the cathode inlet.

The calculated voltage efficiency in these simulations is comparable to those typical of SOFC operation. The temperature differential across the cell is about 30 – 12 °C at the thermal neutral conditions (with average temperature of 800°C), as shown in Fig. 4-a and 4-b respectively. Nevertheless, as the applied potential (and consequently the current density) increases, the temperature gradients and the difference between the maximum and the minimum temperatures in the cell increase. Up to about 350°C of temperature difference is predicted for the cross-flow configuration at relatively high current density (~5000 A/m^2). This is not a desirable condition in the operation of the SOEC since it can cause severe thermal stresses and degradation of the SOEC materials affecting the durability of the system. This indication emphasizes the importance of simulations for performance optimization because operating at higher current density at or above the thermal neutral point with very small temperature gradient can significantly improve the efficiency, durability and cost of the SOEC system.

The simulations also indicate that comparable efficiency at the same $V_{applied}$ can be attained with different flow configurations, but the temperature profiles can be considerably different. Fig. 5 shows a set of examples to this finding. Almost the same current density can be achieved at the given values of $V_{applied}$ for the cross- and parallel-flow configurations. Nevertheless, the parallel-flow SOEC (if achievable in practice) yields lower temperature gradients across the cell, which is favorable for the durability of the cells. The difference is more significant for higher values of $V_{applied}$ and current density.

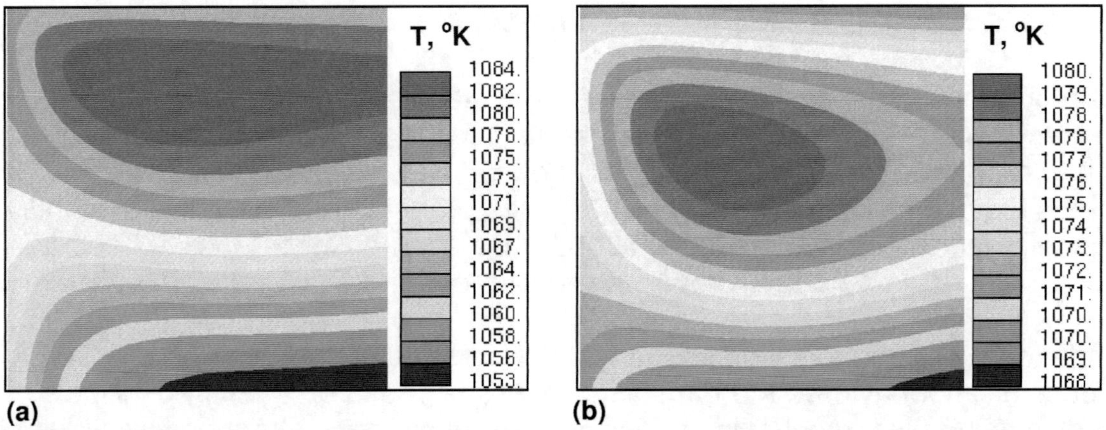

Figure 4. Simulated profiles of temperature in the electrolyte with $T_{average}$ = 800 °C (thermal neutral) (a) 0% mass fraction of H_2 at the cathode inlet ($V_{applied}$:1.125V), and (b) 1.22% mass fraction of H_2 at the cathode inlet ($V_{applied}$:1.169V). Inlet conditions: 800°C and 1atm.

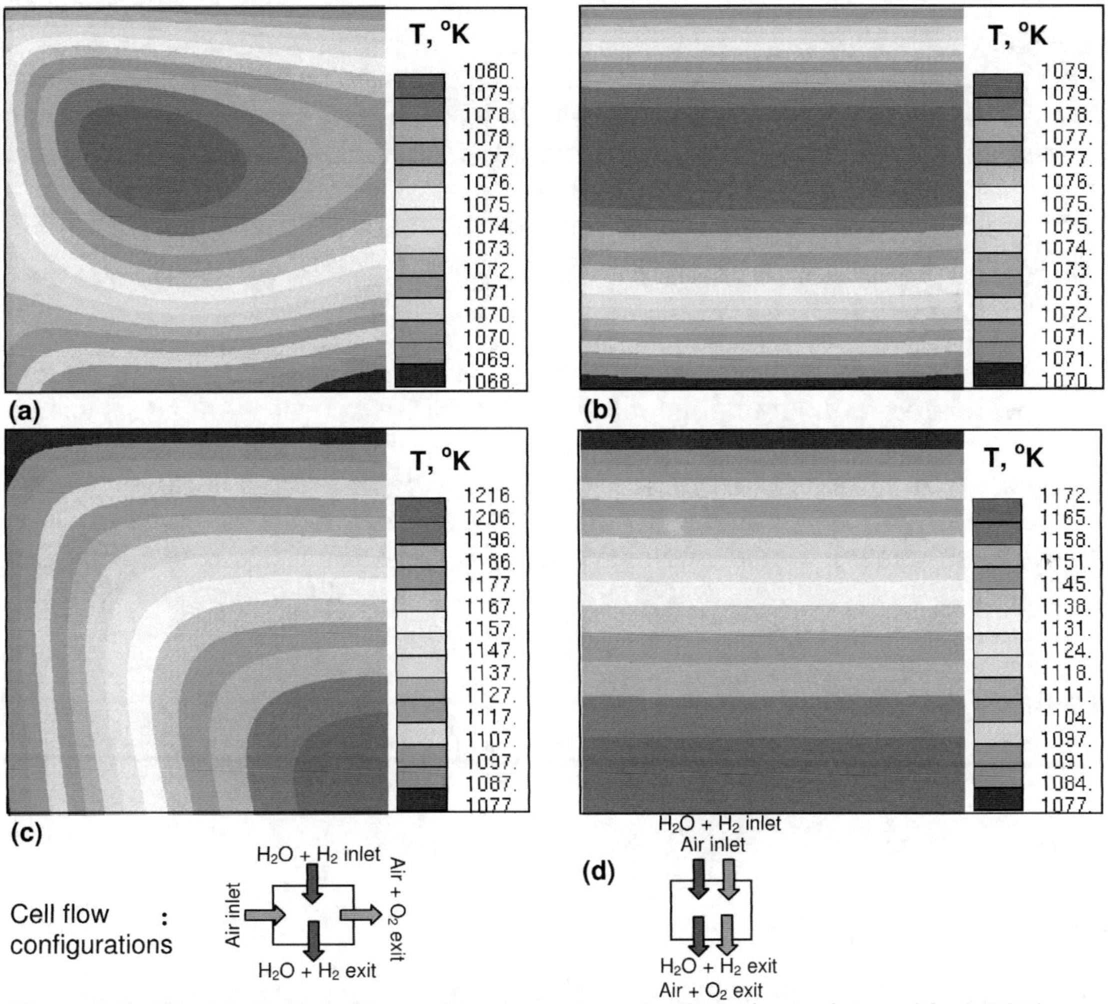

Figure 5. Simulated profiles of temperature in the electrolyte with 1.22% mass fraction of H_2 at the cathode inlet for (a) cross-flow SOEC configuration with $V_{applied}$:1.169V, (b) parallel-flow SOEC configuration with $V_{applied}$:1.169V, (c) cross-flow SOEC configuration with $V_{applied}$:1.25V, (d) parallel-flow SOEC configuration with $V_{applied}$:1.25V. Inlet conditions: 800°C and 1atm.

CONCLUDING REMARKS AND FUTURE WORK

The current simulations do not represent a geometry and flow configuration that minimizes the temperature and current density gradients, but are rather demonstration of the framework for a new capability to predict the coupled electrochemical and CFD behavior of the HTE system. The ultimate objective in the progress of this work is to identify the operating cell and stack flow configurations that can optimize the temperature distribution while maintaining a good hydrogen production performance of the HTE system.

The model developed in this work is the first that has the capability to couple SOEC specific electrochemical behavior to the transport processes based on the first-principals of the governing electrochemical mechanisms. There is yet uncertainty about specifically the reaction activation related phenomena for both the SOFC and SOEC electrodes. Thus, further findings in that area can help improve the mechanism definition in the model presented here. Such an integrated simulation capability can also help eliminate the mechanism related uncertainties that can arise due to using an SOFC specific model with parametric modifications for SOECs. The major differences between the SOFC and SOEC operation of the same cell materials are the reaction activation for catalytic reactions and the thermal behavior due to energetics of the overall reaction. The former is accommodated in this work by including several possible kinetic models for reaction activation controlling mechanism. The latter implies a tendency to decrease the temperature in the cell due to two effects: the heat sink that is associated with the thermal energy required for the electrolysis reaction and the significantly higher heat capacity of the products than that of the reactants at steam electrolysis.

In future work, the current simulation results will be compared to the SOEC polarization and flow related measurements performed at INL. A sensitivity analysis will be carried out to investigate the effects of various assumptions on different key model parameters involved in the calculation of the cell losses and internal heat exchange process. The model will also be extended to other flow configurations to compare their performance to the base case cross-flow configuration. The electrochemical efficiency and uniformity of temperature distribution within the cells will be the major performance parameters for the comparison of the cell and stack configurations.

ACKNOWLEDGEMENTS

We thank the researchers at INL for providing us with the necessary details of the operating conditions and the properties of the SOEC tests for our simulations.

REFERENCES

1. N.F. BESSETTE, W.J. WEPFER, *Journal of Energy Resources Technology,* **117**, 307 (1995).
2. K.P. RECKNAGLE, R.E. WILLIFORD, L.A. CHICK, D.R. RECTOR, M.A. KHALEEL, *Journal of Power Sources*, **113**, 109 (2002).
3. S. CAMPANARI, P. IORA, *Journal of Power Sources*, **132**, 113 (2004).

Flowsheet study of the thermochemical water-splitting IS process for effective hydrogen production

Seiji Kasahara, Shinji Kubo, Ryutaro Hino and Kaoru Onuki*
Department of Advanced Nuclear Heat Technology, Oarai Research Establishment, Japan Atomic Energy Research Institute
Niibori 3607, Narita-cho, Oarai-machi, Higashiibaraki-gun, Ibaraki-ken 311-1394 Japan

Mikihiro Nomura and Shin-ichi Nakao
Department of Chemical System Engineering, Faculty of Engineering, University of Tokyo
7-3-1 Hongo, Bunkyo-ku, Tokyo 113-8656 Japan

*Corresponding author
Phone: +81-29-266-7713
Fax: +81-29-266-7741
E-mail: skasa@popsvr.tokai.jaeri.go.jp

ABSTRACT

Japan Atomic Energy Research Institute (JAERI) has been conducting the research and development on the thermochemical water-splitting IS process for effective hydrogen production using nuclear heat of close to 1000 °C that can be supplied from High Temperature Gas-cooled Reactor (HTGR). The activity covers the studies on the process control for the continuous hydrogen production, the process improvements in the HI decomposition procedure and the preliminary screening of corrosion resistant materials of construction. Present status of the study is presented, especially, focusing on the process flowsheeting study concerning the application of membrane process for the HI processing.

KEYWORDS

hydrogen production, thermochemical water-splitting, IS process, thermal efficiency

INTRODUCTION

Thermochemical water-splitting is the method of obtaining hydrogen from water in thermochemical reaction processes that consist of the combination of endothermic- and exothermic- reactions. This method has a potential advantage of mass production of hydrogen compared with electrochemical water splitting processes because heat can be used directly to produce hydrogen. The concept was first proposed in 1960s by the thermodynamic study [1] and since then over 100 processes have been proposed [2]. The IS (iodine-sulfur, or SI) process is one of the processes that has been researched intensively. This process was proposed in the 70s [3] and it has been studied by, for example, Joint Research Center (JRC)-Ispra [4], General Atomics (GA) [2, 5], Rheinisch Westfälische Technische Hochschule (RWTH) Aachen [6, 7] and Commissariat à l'Energie Atomique (CEA) [8]. The IS process consists of the following reactions

$$I_2 + SO_2 + 2H_2O \rightarrow 2HI + H_2SO_4 \text{ (the Bunsen reaction)}$$

$$H_2SO_4 \rightarrow H_2O + SO_2 + 0.5O_2$$

$$2HI \rightarrow H_2 + I_2$$

Heat of around 900-1000 °C is required for the second reaction. Nuclear heat from High Temperature Gas-cooled Reactor (HTGR) and solar heat are considered as the potential heat sources that emit no CO_2. Japan Atomic Energy Research Institute (JAERI) has been conducting R&D on nuclear hydrogen production using HTGR and the IS process [9, 10]. As for the HTGR technology, the High Temperature Engineering Test Reactor (HTTR), the first HTGR in Japan, achieved reactor-outlet coolant temperature of 950 °C [11]. In this paper, the present status of

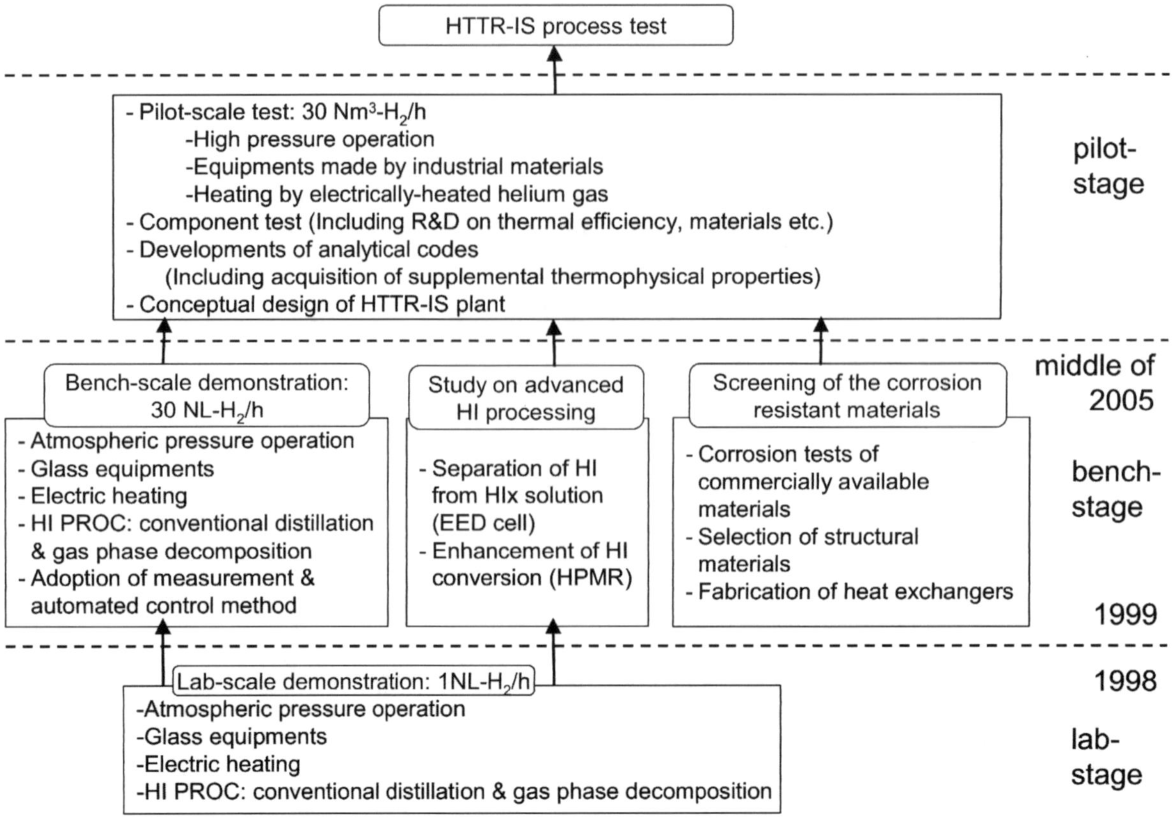

Fig. 1 Overview of the R&D on the IS process at JAERI

study on the IS process is presented focusing on the flowsheet evaluation.

STUDY ON THE IS PROCESS AT JAERI

In JAERI, study on the IS process started in the middle of 80s [12]. Fig. 1 shows the overview of the R&D. In the lab-stage, continuous and stochiometric hydrogen production of 1 NL/h was demonstrated for 48 hours [13]. The result showed the feasibility of the continuous hydrogen production by the IS process.

Bench-scale study is under way in the following three research fields: process control for the long-time stable hydrogen production, HI processing using membrane technologies, and screening of the corrosion resistant materials of construction.

In achieving long-term stable hydrogen production by IS process, an important factor is the stability of compositions of the solutions obtained by the Bunsen reaction. Automated reaction control method and devices were proposed and tested in the bench-scale demonstration experiments with the hydrogen production rate of about 30 NL/h [14]. Fig. 2 shows the simplified schematic diagram of the bench-scale test apparatus. The IS process can be divided into three procedures, namely, the Bunsen reaction procedure (Bunsen PROC), H_2SO_4 decomposition procedure (H_2SO_4 PROC) and HI decomposition procedure (HI PROC). Sulfuric acid and HIx solution (solution of HI-H_2O-I_2) are produced and separated in the Bunsen BROC. In the H_2SO_4 PROC and HI PROC, concentration and/or distillation and thermal decomposition of these acids are carried out. Hydrogen and oxygen are obtained from water through the cycle of the reactions. The other chemicals are recycled in the process. Fig. 3 shows the hydrogen and oxygen production in the experiment. Continuous and stoichiometric operation was carried out for 175 hours. The material balance of water splitting was maintained throughout the experiment. The result verified the effectiveness of the devised control method for long-term stable operation.

Thermal efficiency of hydrogen production of the bench-scale experiment was less than 10 % mainly because of the large heat loss in the HI PROC. Since the HI concentration in the HIx solution obtained in the Bunsen PROC was pseudo-azeotropic or less, HI-H_2O azeotropic mixture of 84 mol% of H_2O was distilled in the distillation column. The vaporization of H_2O resulted in the large heat loss. Also, HI equilibrium conversion ratio is low (22 % at 723 K), which resulted in the large circulation of HI and the heat loss. Therefore, efficient separation of HI from the HIx solution and enhancement of the HI conversion ratio are required to improve thermal efficiency. On these problems, we have been studying an application of membrane technologies. As for the former problem, application of electro-electrodialysis (EED) has been studied to concentrate the HIx solution to over the pseudo-azeotropic composition, which enables to obtain highly concentrated HI in the following distillation. So far, the pseudo-azeotropic HIx solution ([HI] = 10 mol/kg-H_2O) was successfully concentrated to 15 mol/kg-H_2O by beaker-scale experiments [15]. As for the latter, hydrogen permselective membrane reactors (HPMR) equipped with hydrogen permselective membranes have been investigated to improve the HI one-pass conversion ratio by extracting hydrogen from the reaction field. So far, HI conversion enhancement has been demonstrated using a silica membrane prepared by chemical vapor deposition method [16].

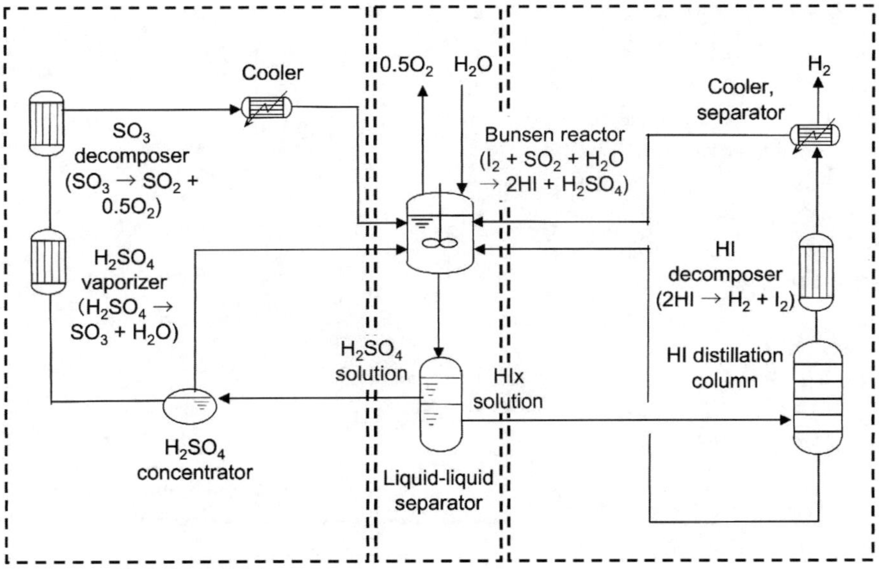

Fig. 2 Simplified schematic diagram of the bench-scale test apparatus

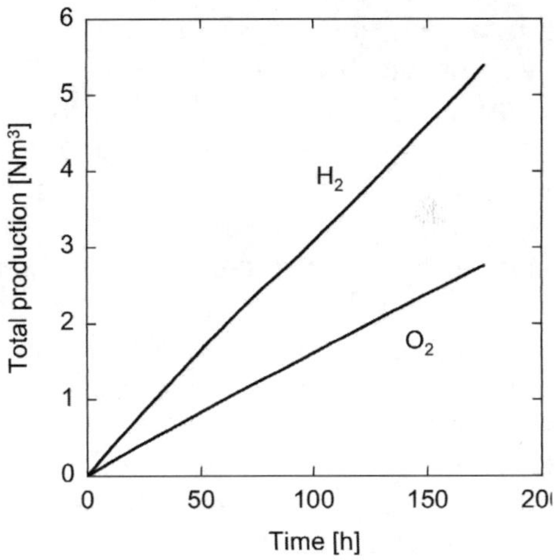

Fig. 3 Hydrogen and oxygen production of the bench-scale demonstration experiment

Corrosive chemicals such as H_2SO_4, HI and I_2 are used at high temperatures in the IS process, and the corrosion resistant materials are necessary. Selection of the materials by corrosion tests has been carried out [17-19]. Fig. 4 summarizes the candidate materials. Refractory alloys such as Hastelloy showed a certain corrosion resistance in the gas phase environments. On the other hand, in the liquid phase environments, only exotic materials, e.g., glass, ceramics and rare metals, showed the corrosion resistance. Because ceramics are brittle and rare metals are expensive, fabrication of apparatuses are under study where these materials are used only for the parts that contact with the process fluids. Concerning the H_2SO_4 vaporizer, a prototype of heat exchanging block made of Si-SiC was fabricated [20]. Fig. 5 shows the upper half of the block. Sulfuric acid flows up inside the block and is vaporized in the block. The heat of vaporization is supplied by high temperature helium gas that flows in the block through the holes at the side of the block and flows down. Structural integrity of the block was verified by thermal stress analysis.

It is planned that the bench-scale study is to be completed in the middle of 2005, and the pilot test follows. Pilot-scale water splitting test will be conducted with hydrogen production rate of about 30 Nm^3/h. The test plant will be made of industrial materials and operated under elevated pressure conditions. The process heat will be supplied by electrically-heated helium gas. Development of analytical codes, conceptual design of the HTTR-IS plant will be carried out, as well.

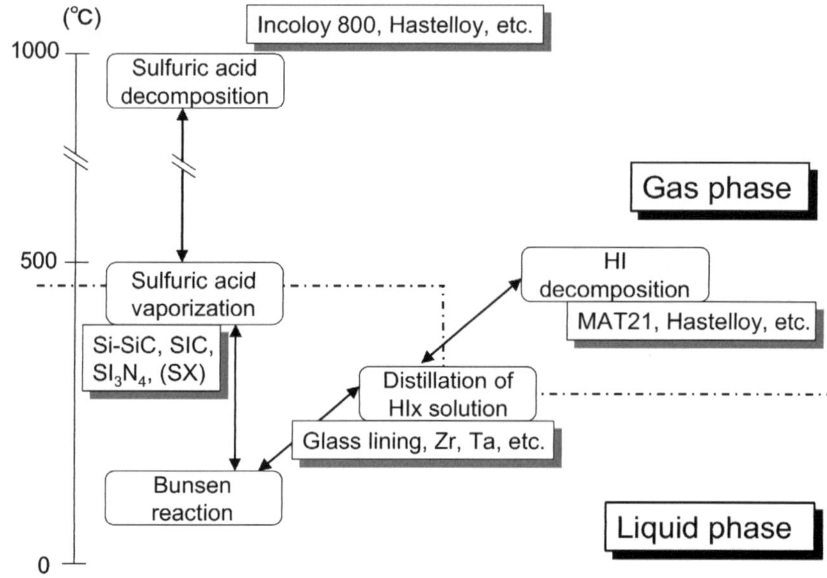

(SX and MAT21 are the brand names of a high-silicon stainless steel alloy (Edmeston AB/Sandvik AB) and a Ni-Cr-Mo-Ta alloy (Mitsubishi Materials Corp), respectively.)

Fig. 4 Candidate materials of construction

ANALYSIS OF A FLOWSHEET FEATURING MEMBRANE TECHNOLOGIES

Thermal efficiency of hydrogen production by the IS process is influenced by the selection of apparatus. Norman et al. [5] first reported the thermal efficiency of 47 % based on an analysis of a flowsheet that featured extractive distillation of HIx solution using phosphoric acid and multi-stage flash evaporation for concentration of sulfuric acid. Brown et al. [2] and Goldstein et al. [8] investigated heat/mass balance of the flowsheet featuring reactive distillation of HIx solution [6]. There seems to be a space for further study especially in the HI PROC considering the use of another chemical in extractive distillation scheme and the very corrosive environments in the reactive distillation scheme. Therefore, in JAERI, the process featuring the membrane technologies as stated in the previous section has been studied. The followings summarize the present status of flowsheet study in JAERI [21-24].

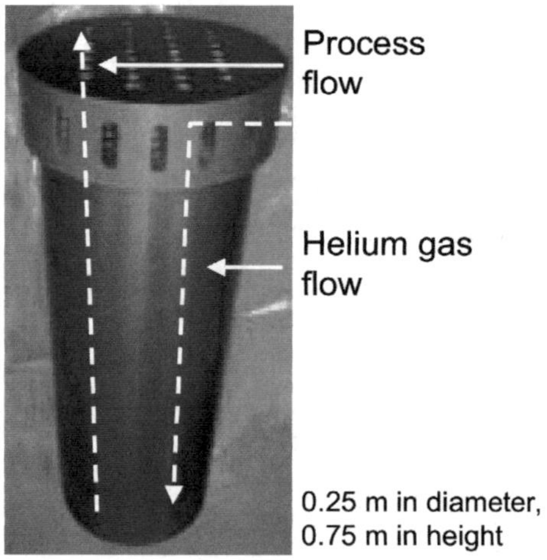

Fig. 5 Prototype of heat exchanging block made of Si-SiC for H_2SO_4 vaporizer

Fig. 6 illustrates the schematic flowsheet of the membrane process. The HIx solution from the Bunsen PROC is, at first, fed to the EED cell in order to increase the HI concentration to over pseudo-azeotropic one. The concentrated HIx solution is then fed to the HI distillation column. In the HI distillation, HI-rich vapor is obtained from the top and HIx solution of pseudo-azeotropic composition is obtained at the bottom. The HI-rich vapor is fed to the HI decomposition reactor equipped with a hydrogen permselective membrane (HPMR), where only H_2 is separated from the reaction mixture owing to the selective permeation through the membrane. Unreacted HI, remaining H_2 and I_2 are then cooled down to make partial condensation of I_2. The liquid I_2 is separated at the HI-I_2 separator and returned to the Bunsen PROC. The rest is recycled to the HPMR. In the analysis, sulfuric acid with molar ratio of H_2SO_4 : H_2O =1 : 4.11 and HIx solution with molar ratio of HI : H_2O : I_2 =1 : 5.3 : 3.8 are assumed to be obtained from the Bunsen PROC [5]. As for H_2SO_4 PROC, heat/mass balance reported by Knoche et al. [7] was used which applied multiple-effect evaporators for the concentration of sulfuric acid.

Thermal efficiency (η) was calculated by the equation (1).

$$\eta = \frac{\text{Higher heating value of hydrogen (285.8 kJ/mol)}}{H_{heat} + H_{elec.}} \times 100 \qquad (1)$$

Here, H_{heat} denotes the net heat demand and $H_{elec.}$ denotes the heat required to produce the net electricity demand. In the heat balance calculation, following assumptions were applied:

- Internal heat exchange is carried out as much as possible.
- The process is operated adiabatically.
- Work required in pumping is not counted.
- Electricity generation efficiency for producing the net electricity demand is 40 %.

Details of the calculation are reported elsewhere [21, 22].

At first, upper bound of thermal efficiency of the membrane process (case 1) was evaluated using the operation parameters shown in Table 1. Here, overpotential of the EED cell ($\triangle V_{EED}$) was set at 0 V. This corresponds to the assumption that the EED cell could be operated without electric resistance. As for apparent transport number of proton (t_+) and electroosmosis coefficient (β) of the cation exchange membrane (CEM) used in the EED cell, experimental data [24] were used. Temperature difference at heat exchangers ($\triangle T_{exc.}$) was set at 0 K. The heat exchangers were assumed to transfer heat with no heat loss. Also, it was assumed that power generation using the low temperature waste heat was carried out with the electric recovery ratio from waste heat ($\eta_{rec.}$) of 15 %. As for the HI one-pass conversion ratio, it was set at 50 % considering the experimentally obtained figure of 61.3 % [16]. The upper bound of thermal efficiency was estimated to be 57 % [21]. However, when electric resistance at the EED cell ($\triangle V_{EED}$ of 0.2 V [15]) and heat loss at the heat exchangers ($\triangle T_{exc.}$ of 10 K) were considered and no electricity recovery was assumed ($\eta_{rec.}$ of 0 %) (case 2), η of 34 % was evaluated [22].

Next, the effect of operation parameters on the thermal efficiency was carried out. As for the EED, sensitivities of the following parameter were examined: ratio of mass flow rates of flow No. 8 and that of No. 7, apparent transport number of proton of the CEM (t_+), apparent electroosmosis coefficient of the CEM (β), overpotential ($\triangle V_{EED}$) and concentration of HI at cell exit (flow No.10) (m_{EED}). As for the HI distillation column, sensitivities of the operation pressure and the reflux ratio were evaluated. The effect of HI one-pass conversion ratio at the HPMR, and that of $\triangle T_{exc.}$ were also investigated. Table 2(a) shows the results of the analysis [22]. Here, the heat/mass balance calculation was carried out using the figures of operation parameters assumed in the case 2, and the calculated sensitivities are expressed in terms of the relative changing ratio of parameters that were required to change η by 1 %. The operation parameters relating to the EED cell was found to influence the thermal efficiency largely. Especially, m_{EED} had the greatest influence. The results show the R&D on the EED cell is most effective to attain high thermal efficiency. For example, as shown in Table 1, η of over 40 % could be achieved by the improvement of property of the

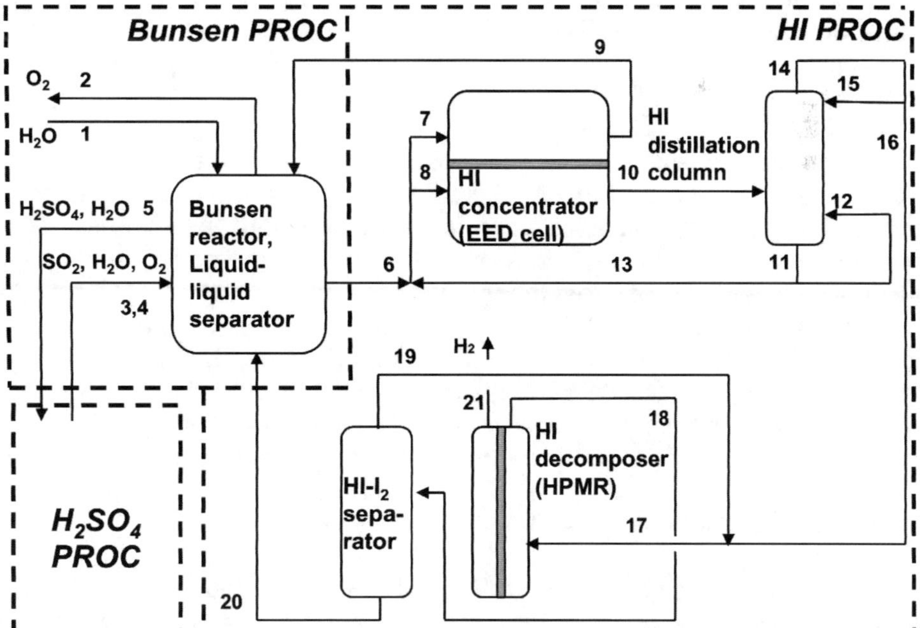

Fig. 6 Schematic flowsheet of the membrane process [22]

Table 1 Operation parameters of HI PROC

Apparatus	Operation parameter	Unit	Parameters for upper bound thermal efficiency (case 1) [21]	Parameters considering the losses (case 2) [22]	Example of parameter set for η of over 40 % (case 3)
EED cell	ratio of mass flow rates of flow No. 8 to that of No. 7	-	1	0.5	0.5
	apparent transport number of proton of the CEM (t_+)	mol/mol-e^-	0.8	0.8	0.98
	apparent electroosmosis coefficient of the CEM (β)	mol/mol-e^-	1.2	1.2	1.1
	overpotential (ΔV_{EED})	V	0	0.2	0.15
	concentration of HI at cell exit (flow No.10) (m_{EED})	mol/kg-H_2O	11.7	13.5	13.5
HI distillation column	pressure in the column	$\times 10^5$ Pa	11.7	20.7	23.1
	reflux ratio at the column	-	0.5	1	0.5
	temperature at the bottom	K	485	522	522
	temperature at the top	K	316.4	341.72	347.13
HPMR for HI decomposition	HI one-pass conversion ratio	%	50	50	50
Heat exchanger	temperature difference ($\Delta T_{exc.}$)	K	0	10	10
Turbine for waste heat recovery	electric recovery ratio from waste heat ($\eta_{rec.}$)	%	15	0	0
Thermal efficiency (η)		%	57	34	41

CEM (t_+, β) and cell performance (ΔV_{EED}) (case 3).

Sensitivity was also analyzed of the chemical properties relating to the HIx solutions, on which accurate data were not obtained yet [22]. Table 2(b) shows the results. The mole fraction of HI and I_2 in the HIx solution at the inlet of the HI PROC (flow No. 6) and the HI concentration of pseudo-azeotropic HIx solution had great effects on the thermal efficiency. Caloric properties such as dilution enthalpy and heat capacity showed smaller effects. The higher thermal efficiency was obtained at the higher mole fraction of HI at the inlet of HI PROC [22]. Therefore, it is important to obtain HI-rich HIx solution in the Bunsen PROC to attain higher thermal efficiency. The result clarified that the accurate VLT data of HIx solution is also very important.

In the above-mentioned flowsheet study, H_2SO_4 PROC was fixed at constant condition. Flowsheet study of the H_2SO_4 PROC is under way.

SUMMARY

The present status of R&D at JAERI on the IS process was briefly described. Continuous and stable operation of the bench-scale apparatus was successfully carried out. Membrane techniques aimed for the application in HI PROC were investigated. Preliminary screening the corrosion resistant materials of construction has been performed, and a ceramic heat exchanger for H_2SO_4 vaporizer has been test-fabricated.

Flowsheet study has been carried out on the membrane application for HI processing. Upper bound thermal efficiency was estimated to be 57 %. Sensitivity analysis of operation parameters indicated the performance of the EED cell was very important. Also, it is effective for improving the heat balance to attain HI-rich HIx solution in the Bunsen reaction. The acquisition of accurate VLE data of HIx solutions was important for accurate evaluation of the thermal efficiency.

AKNOWLEDGEMENTS

The bench-scale demonstration, and the test-fabrication of heat exchanger for H_2SO_4 vaporizer have been carried out under the contract of research between JAERI and the Ministry of Education, Culture, Sports, Science and Technology of Japan.

Table 2(a) Sensitivity analysis of operation parameters in HI PROC in case 2

Apparatus	Operation parameter	Unit	Relative changing ratio of operation parameter to decrease η by 1 % [%]	Order of sensitivity
EED cell	ratio of mass flow rates of flow No. 8 to that of No. 7	-	22	5
	apparent transport number of proton of the CEM (t_+)	mol/mol-e^-	6.3	2
	apparent electroosmosis coefficient of the CEM (β)	mol/mol-e^-	16.7	4
	overpotential ($\varDelta V_{EED}$)	V	15.5	3
	concentration of HI at cell exit (flow No.10) (m_{EED})	mol/kg-H_2O	5.0	1
HI distillation column	pressure in the column	$\times 10^5$ Pa	59.1	7
	reflux ratio at the top of the column	-	126	8
HPMR for HI decomposition	HI one-pass conversion ratio	%	*	9
Heat exchanger	temperature difference ($\varDelta T_{exc.}$)	K	53	6

*η was not changed by 1 % within the range of 0-100 %.

Table 2(b) Sensitivity analysis of chemical properties in HI PROC in case 2

Chemical property	Unit	Relative changing ratio of chemical property to decrease η by 1 % [%]	Order of sensitivity
activities of HI in solution in the EED cell	-	60	5
heat capacity of HI in HIx solution	kJ/(K·mol)	**	7
heat capacity of I_2 in HIx solution	kJ/(K·mol)	228	6
dilution enthalpy of HI in HIx solution	kJ/mol	23	4
mole fraction of HI in the HIx solution at the inlet of HI PROC. (flow No.6)	mol%	4.0	1
mole fraction of I_2 in the HIx solution at the inlet of HI PROC. (flow No.6)	mol%	8.2	3
HI concentration of pseudo-azeotropic HIx solution	mol%	4.7	2

**decrease of η was less than 1 % even when the decrease of the property was 100 %.

REFERENCES

[1] J. E. Funk and R. M. Reinstrom, *I&EC Process Design and Develop.*, 5, 336-342 (1966).

[2] L. C. Brown, G. E. Besenbruch, R. D. Lentsch, K. R. Schultz, J. F. Funk, P. S. Pickard, A. C. Marshall and S. K. Showalter, *GA-A24285* (2003).

[3] J. L. Russell, Jr., K. H. McCorkle, J. H. Norman, J. T. Porter II, T. S. Roemer, J. R. Schuster, and R. S. Sharp, *Proc. of 1st World Hydrogen Energy Conf.*, Miami Beach, U. S. A., 1A105-1A124 (1976).

[4] G. De Beni, G. Pierini, B. Spelta, *Int. J. Hydrogen Energy*, 5, 141-149 (1980).

[5] J. H. Norman, G. E. Besenbruch, L. C. Brown, D. R. O'Keefe and C. L. Allen, *GA-A16713* (1982).

[6] M. Roth and K. F. Knoche, *Int. J. Hydrogen Energy*, 14, 8, 545-549 (1989).

[7] K. F. Knoche, H. Schepers and K. Hesselmann, *Proc. of 5th World Hydrogen Energy Conf.*, July 15-20, Toronto, Canada, 487-502 (1984).

[8] S. Goldstein, J.-M. Borgard, and X. Vitart, *Int. J. Hydrogen Energy*, 30, 619-626 (2005).

[9] K. Onuki, H. Nakajima, I. Ioka, M. Futakawa and S. Shimizu, *JAERI-Review* 94-006 (1994).

[10] K. Onuki, S. Kubo, S. Kasahara, S. Ishiyama, H. Nakajima, S. Higashi, T. Masaki and T. Takeda, *Proc. of 12th Int. Conf. Nucl. Eng.,* ICONE12-49364, Arlington, Virginia USA, April 25-29, (2004).

[11] S. Fujikawa, H. Hayashi, T. Nakazawa, K. Kawasaki, T. Iyoku, S. Nakagawa and N. Sakaba, *J. Nucl. Sci. Technol.*, 41, 12, 1245-1254 (2004).

[12] S. Shimizu, *International Symposium on Advanced Nuclear Energy Research —Near-future chemistry in nuclear energy field—*, 324-331, Oarai, Ibaraki, Japan, February 15-16 (1989).
[13] H. Nakajima, M. Sakurai, K. Ikenoya, G.-J. Hwang, K. Onuki, and S. Shimizu, *Proc. of 7th Int. Conf. Nucl. Eng.,* ICONE-7104, Tokyo, Japan, April 19-23, (1999).
[14] S. Kubo, H. Nakajima, S. Kasahara, S. Higashi, T. Masaki, H. Abe, K. Onuki, *Nucl. Eng. Des.*, 233, 347-354 (2004).
[15] G.-J. Hwang, K. Onuki, M. Nomura, S. Kasahara, and J.-W. Kim, *J. Membr. Sci.*, 220, 129-136 (2003).
[16] M. Nomura, S. Kasahara and S. Nakao, *Ind. Eng. Chem. Res.,* 43 (18), 5874-5879 (2004).
[17] M. Futakawa, S. Kubo, T. Wakui, K. Onuki, S. Shimizu and A. Yamaguchi, *Jikken Rikigaku*, 3, 2, 109-114 (2003) [in Japanese].
[18] S. Kubo, M. Futakawa, I. Ioka, K. Onuki, S. Shimizu, K. Ohsaka, A. Yamaguchi, R. Tsukada, and T. Goto, *Proc. of 2003 AIChE Spring National Meeting*, March 30 – April 3, New Orleans, Louisiana, U. S. A., (2003).
[19] N. Nishiyama, M. Futakawa, I. Ioka, K. Onuki, S. Shimizu, M. Eto, T. Oku and M. Kurabe, *J. Soc. Mat. Sci. Japan*, 48, 7, 746-752 (1999) [in Japanese].
[20] H. Ota, S. Kasahara, A. Terada, S. Kubo, K. Onuki, R. Hino, T. Inatomi, K. Ogura, M. Kobayashi, M. Hodotsuka, S. Maruyama, submitted to *13th Int. Conf. Nucl. Eng.*, ICONE13-50494, May 16-20, Beijing, China, (2005).
[21] S. Kasahara, G.-J. Hwang, H. Nakajima, H.-S. Choi, K. Onuki and M. Nomura, *J. Chem. Eng. Japan*, 36, 7, 887-899 (2003).
[22] S. Kasahara, S. Kubo, K. Onuki, M. Nomura and S. Nakao, *Proc. of 15th World Hydrogen Energy Conf.*, 29J-03, Yokohama, Japan (2004).
[23] S. Kasahara, S. Kubo, K. Onuki and M. Nomura, *Int. J. Hydrogen Energy,* 29, 579-587 (2004)
[24] K. Onuki, G.-J. Hwang, Arifal, S. Shimizu, *J. Membr. Sci.,* 192, 193-199 (2001).

NOMENCLATURE

H_{heat}	net heat demand of the IS process from heat source	[kJ/mol-H_2]
$H_{elec.}$	heat required for the net electricity demand	[kJ/mol-H_2]
m_{EED}	concentration of HI at the exit of the cell (flow No. 10)	[mol/kg-H_2O]
$t+$	apparent transport number of proton of CEM in the EED cell	[mol/mol-e^-]
β	apparent electroosmosis coefficient of CEM in the EED cell	[mol/mol-e^-]
$\Delta T_{exc.}$	temperature difference at heat exchangers	[K]
ΔV_{EED}	overpotential in the EED cell	[V]
η	thermal efficiency	[%]
$\eta_{rec.}$	electric recovery ratio from waste heat	[%]

H_2/Methanol Production by Sulfur-Iodine Thermochemical Cycle Powered by Combined Solar/Fossil Energy

*A. Giaconia, R. Grena, M. Lanchi, R. Liberatore, P. Tarquini**

ENEA Research Center "Casaccia" via Anguillarese, 301 – 00060 S. M. di Galeria, Rome (Italy).
** corresponding author: tarquini@casaccia.enea.it; fax: +39 06 30486779, tel: +39 06 30484539.*

Abstract

Hydrogen production from water using sulfur-iodine (S-I) thermochemical cycle, powered by combined solar and fossil heat sources, has been investigated in terms of process efficiency and costs. The combined energy supply has been conceived in order to operate the chemical process continuously: a solar concentrator plant supplies thermal load for services at medium temperatures (<550°C) by the use of molten salts as energy carrier with a large scale heat storage, while a fossil fuel furnace provides heat load for endothermic reactor operating at higher temperatures. This device permits the plant to provide continuously thermal power at a constant rate, regardless of variations in solar power availability. Moreover, the S-I chemical process has been interfaced to a methanol production plant fed with the carbon oxides generated by fossil fuel combustion and a fraction of the hydrogen produced. Since the sulfuric acid concentration/decomposition section of the chemical plant is interfaced to both the fossil fuel furnace and the solar plant, it has been studied more exhaustively with process simulations and plant sizing. The energetic efficiency of the whole chemical process is about 34.5%, where much of the total energy demand (ca. 70%) is renewable. An industrial scale plant with hydrogen capacity of about 26,000 tons/year, coupled to a solar plant with an average thermal power of about 267 MW_{th} has been considered, and the specific cost of the produced hydrogen and methanol has been determined.

1. Introduction

The worldwide increasing energy demand together with the environmental alarms due to greenhouse gas emissions (designated as GHG) have promoted research work on the utilization of renewable energy sources. On this perspective large scale hydrogen production from water supported by solar energy is an appealing route.

There are still many questions to answer for final introduction of hydrogen as large scale energy carrier, regarding both hydrogen production, with the development of large scale low-cost sustainable processes, and utilization, with the design of high efficiency and low-cost fuel cells and the set up of safe hydrogen storage systems with high capacity.

At present large scale and low-cost hydrogen production is attained from fossil fuel charge, mainly by steam reforming of natural gas, leading to the production of about 50 million tons/year of hydrogen (Lipman T.E., 2004), with a demand that is expected to rapidly increase in the next few decades. This conventional route has the drawback of the increasing cost and low availability of charge for long term and, from the environmental point of view, the CO_2 emission, when the general trend is the GHG reduction.

The pathway to hydrogen generation from entirely renewable energy sources probably goes by a transitional period with the utilization of hybrid fossil/renewable integrated systems. An example is the development of solar steam reforming or coal gasification processes (Kodama T., 2003).

The most innovative technologies for large scale hydrogen production from water are essentially high temperature processes, given that the only available low temperature process is the electrolysis of water solutions, which is not considered a large scale process owing to its low efficiency. These high temperature processes include pure thermochemical cycles (e.g. sulfur-iodine, UT-3, etc.), hybrid electrochemical-thermochemical cycles (e.g. Westinghouse) and advanced electrolysis.

One of the most promising thermochemical cycles is the sulfur-iodine (S-I) cycle, owing to the relatively low operating temperatures and the presence of fluid phases (gas and liquid). It basically consists of the following chemical reactions, carried out at three different temperature levels (low, medium and high temperature):

- Bunsen reaction (exothermic at 20-120°C):
  $2H_2O + I_2 + SO_2 \rightarrow H_2SO_4 + 2HI$
- Hydriodic acid decomposition (endothermic, at 300-450°C):
  $2HI \rightarrow I_2 + H_2$
- Sulfuric acid decomposition (endothermic, at 800-900°C):
  $H_2SO_4 \rightarrow H_2O + SO_3 \rightarrow H_2O + SO_2 + \frac{1}{2} O_2$

Here we present a new continuous process for hydrogen generation from water based on the sulfur-iodine cycle powered by a mixed solar/fossil heat source. This combined energy supply system has been conceived in order to operate the chemical process continuously: a solar concentrator plant supplies thermal load at low-medium temperatures (< 550°C), while a methane furnace provides heat load for the endothermic reactor operating at higher temperatures, i.e. the sulfuric acid decomposition reaction.

In order to take advantage of the carbon oxides generated by methane combustion, we thought right to convert them to methanol in a methanol section fed with part of the hydrogen generated: as known methanol is a high added value commodity and a very useful energy carrier with lower CO_2 emissions than common liquid fuels and more suitable than hydrogen for transportation, storage and handling (Xin Dong, 2003).

2. Process description

The simplified scheme of the process with mass balances is illustrated in figure 1.

In the Bunsen section sulfuric and hydriodic acids are generated and separated. The former is sent to the "sulfuric section" where it's concentrated and dissociated into sulfur dioxide and oxygen, both recycled to the Bunsen section.

The hydriodic acid (HI) from Bunsen section is fed to the "HI_x section" where it is concentrated and finally decomposed to hydrogen and iodine: iodine is recycled back to the Bunsen reactor, while hydrogen is partially fed to the methanol section.

As abovementioned, the heat supply for the endothermic reactions is provided by two different systems:

- a solar collection plant, where molten salts at temperatures up to 550°C are used as thermal energy vector and heat storage; such molten salts have an excellent behavior between 290-600 °C but, as their utilization at lower temperatures is precluded owing to the increase of viscosity and solidification (melting point is about 230°C), Dowtherm HT has been chosen as thermal fluid for heat duties between 300 and 90 °C.

- a methane furnace as heat source at higher temperatures. The oxygen produced in the thermochemical cycle is fed to the methane furnace in order to maximize the efficiency of the fossil fuel combustion.

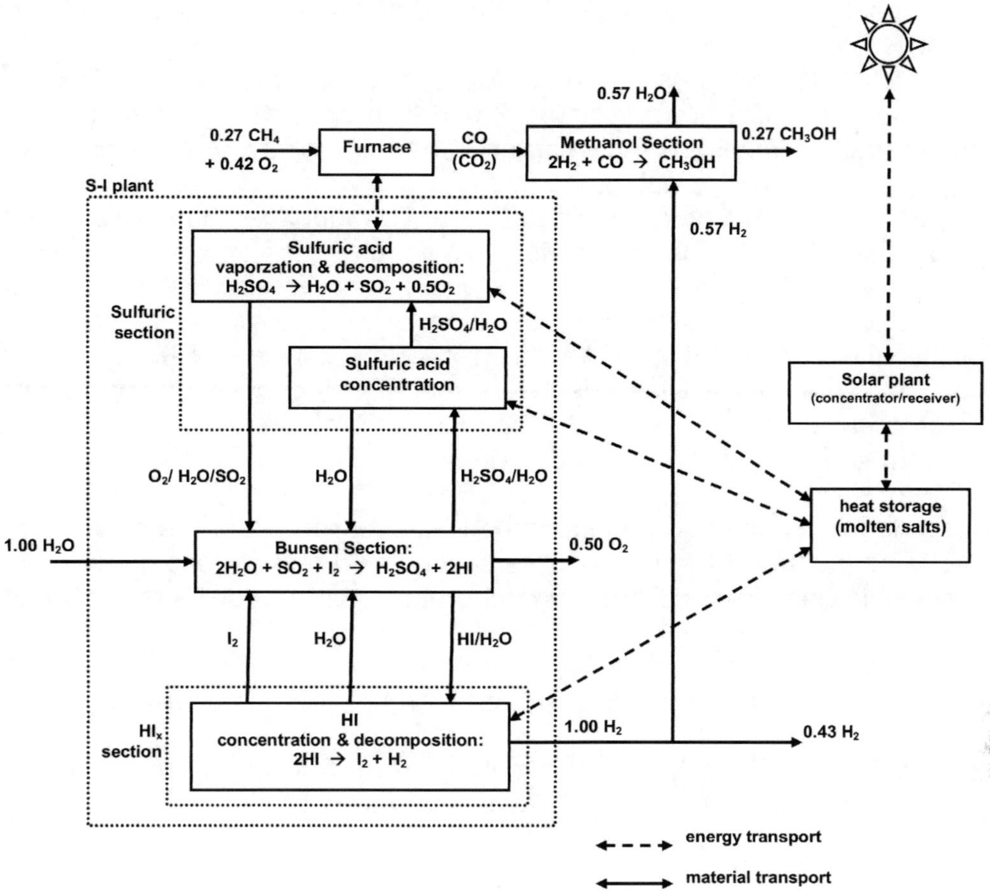

Figure 1. General scheme for the Sulfur-Iodine process for hydrogen/methanol production powered by combined solar/fossil sources.

2.1. Thermochemical Process

The general scheme of the chemical plant is essentially constituted by three interconnected sections (sulfuric, Bunsen, HI_x), with an additional section for methanol synthesis as schematized in figure 1.

Since the sulfuric section is interfaced to both the methanol section (through the methane furnace) and the solar plant (by means of the molten salts as thermal energy carrier), it was studied more exhaustively, with the development and optimization of a flowsheet. A complete flowsheet for the methanol section was developed too. Concerning the Bunsen and HI_x sections, the process scheme proposed earlier by General Atomics (Brown L. C., 2003) was assumed.

The process simulation program selected for material and heat balances is ASPEN Plus from Aspen Technology.

2.1.1. Sulfuric Section

The sulfuric acid section has been extensively studied in terms of flowsheet analysis, thermodynamic yield optimization and equipment sizing; where possible the stream matching with heat recovery has been employed in order to reduce the overall heat consumptions.

For clarity, this section has been divided into sulfuric acid concentration and decomposition subsections, shown in figure 2 and 3, respectively. The mass balance and composition of streams labeled in the figures 2 and 3 are given in tables 1 and 2 with reference to the production of 1 mol SO_2.

Regarding thermodynamic modeling, the ElecNRTL model has been used, as proposed by GA (Brown L.C., 2003). This model is an electrolytic variation of the NRTL model, involving the description of the ionic associations and dissociations along with pure component and pair-wise parameters for the ionic and non-ionic components. Moreover, because of the high concentration of the sulfuric acid solution and the high temperature required for evaporation, the existing ElecNRTL model with conventional parameters is not adequate to properly represent this peculiar sulfuric system. Therefore a specific model with optimized parameters set (provided and regressed by Aspen Technologies) has been applied.

Concerning the sulfuric acid concentration subsection, the main purpose of the conceived scheme (figure 2) is to concentrate sulfuric acid before decomposition in order to reduce the sensible heat demand at high temperatures in the downstream decomposition subsection.

The sulfuric acid concentration has been achieved first from 57 wt.% to 81 wt.% in four evaporators (B101, B102, B103 and B104), and then up to 90 wt.% in a distillation column (C101).

The heat required in this subsection is entirely provided by the solar plant facility by means of the molten salts/Dowtherm HT heat exchanger E110.

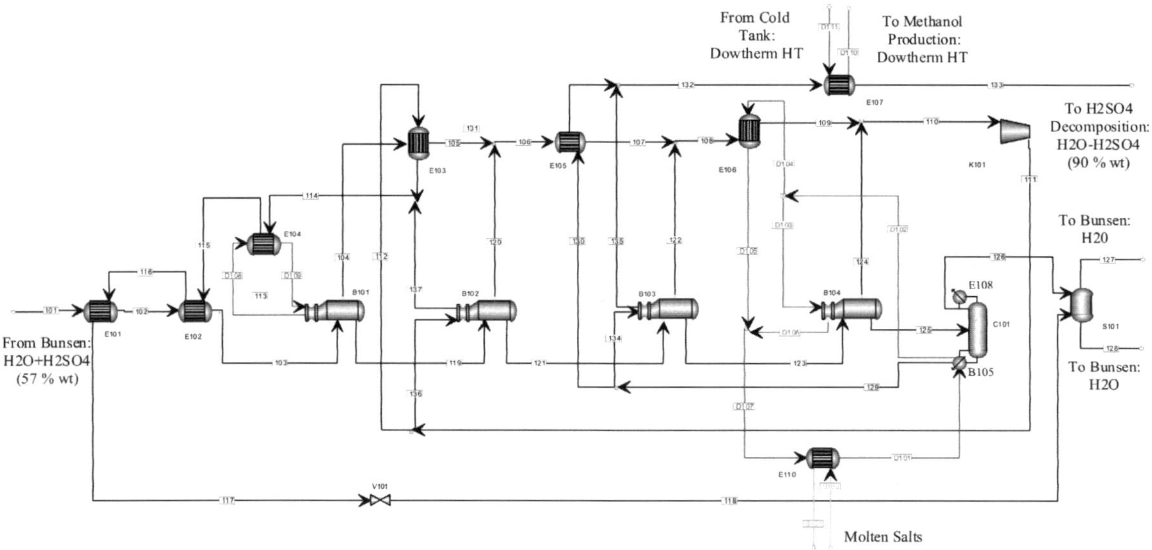

Figure 2. Flowsheet for the sulphuric acid concentration subsection.

Table 1. Mass balance and stream composition for the sulfuric acid concentration subsection (1 mol SO_2 production).

Stream	Mole flow (mol/mol_{SO2})					T (°C)	P (atm)	vapour fraction
	H_2O	H_2SO_4	SO_2	SO_3	O_2			
101	4.19	1.02				120	1	0.00
102	4.19	1.02				126	1	0.00
103	4.19	1.02				134	1	0.00
104	2.10	(< 0.01)				171	1	1.00
105	2.10	"				188	1	1.00
106	2.53	"				188	1	1.00
107	2.53	"				197	1	1.00
108	2.72	"				197	1	1.00
109	2.72	"				205	1	1.00
110	2.86	"				205	1	1.00
111	2.86	"				451	4	1.00
112	0.17	"				451	4	1.00
113	0.17	"				186	4	1.00
114	2.86	"				186	4	1.00
115	2.86	"				144	4	0.03
116	2.86	"				144	4	0.00
117	2.86	"				132	4	0.00
118	2.86	"				100	1	0.00
119	2.09	1.02				171	1	0.00
120	0.43	(< 0.01)				188	1	1.00
121	1.66	1.02				188	1	0.00
122	0.19	(< 0.01)				197	1	1.00
123	1.47	1.02				197	1	0.00
124	0.15	(< 0.01)				205	1	1.00
125	1.33	1.02				205	1	0.00
126	0.70	(< 0.01)				100	1	0.00
127	0.17					100	1	1.00
128	3.40	(< 0.01)				100	1	0.00
129	0.62	1.02				257	1	0.00
130	0.04	0.06				257	1	0.00
131	0.04	0.06				200	1	0.00
132	0.62	1.02				200	1	0.00
133	0.62	1.02				105	1	0.00
134	0.58	0.95				257	1	0.00
135	0.58	0.95				200	1	0.00
136	2.70	(< 0.01)				451	4	1.00
137	2.70	(< 0.01)				186	4	1.00

The sulfuric acid decomposition subsection (figure 3) is linked to the concentration unit by means of sulfuric acid storage at atmospheric pressure and ambient temperature. The concentrated sulfuric acid (90 wt.% H_2SO_4) from the concentration subsection is pumped at 6 atm with pump P201 into the SO_3 adsorption tower C201, where it is heated up to 310°C and contacted with a countercurrent flow of incoming gaseous product from the decomposition reactor (stream 209), consisting of water, non-decomposed SO_3 and H_2SO_4 (as reported in table 2). The operative pressure selected for the whole unit (6 atm) represents the compromise value for efficient operation of the column C201, which requires high pressure, and high SO_2 yield in the decomposition reactor R201, which is improved at lower pressures.

Liquid sulfuric acid at 310°C from the bottom of the separation tower C201 is sent to the first step of the reaction sequence consisting of sulfuric acid evaporation and partial decomposition into SO_3 and water, performed in the boiler B201 and in the superheater E201, both powered by the sensible heat of molten salts. Stream 205 (at 520°C) exiting exchanger E201 is sent to heat recuperator E202, where it is heated up to 640°C by the heat of product gases from reactor R201. After, stream 205A enters the methane furnace F201, where the gas temperature is increased up to 1095°C and the thermal decomposition of SO_3 partially occurs. However, most part of SO_3 conversion is carried out in the catalytic reactor R201 operating adiabatically. The total amount of energy required for the catalytic decomposition is supplied by the methane furnace as sensible heat of the combustion products. At the fixed operative conditions of 850°C and 6 atm, the equilibrium conversion of SO_3 to SO_2 is about 70%.

Streams 213 and 215 (consisting of SO_2 and water), are finally sent to the Bunsen section while the carbon oxides from methane are sent to the methanol section (stream 303).

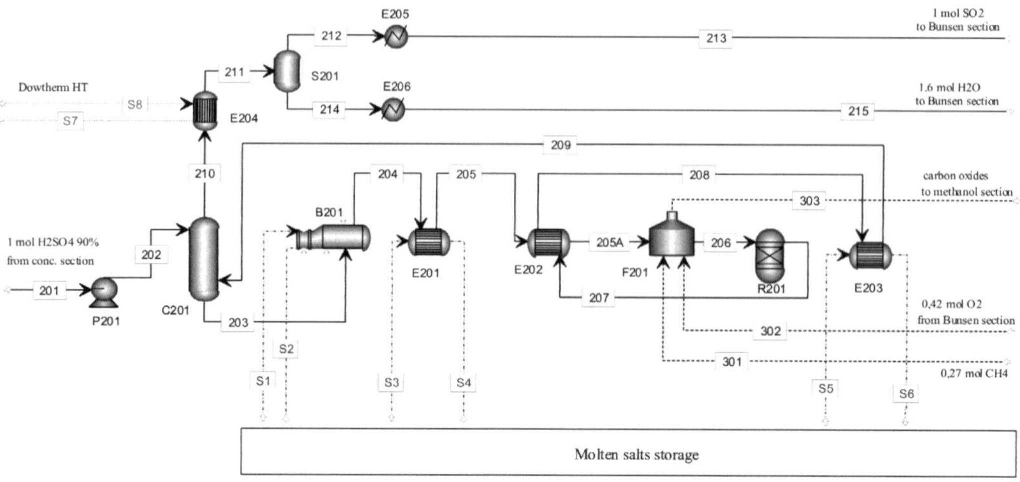

Figure 3. Flowsheet for the sulphuric acid decomposition subsection

Table 2. Mass balance and stream composition for the sulfuric acid decomposition subsection (1 mol SO_2 production).

Stream	Mole flow (mol/mol_{SO2})					T (°C)	P (atm)	vapour fraction
	H_2O	H_2SO_4	SO_2	SO_3	O_2			
201	0.61	1.00				25	1	0.00
202	0.61	1.00				26	6	0.00
203	0.82	1.43				311	6	0.00
204	0.82	1.43				417	6	1.00
205	1.97	0.28		1.15		520	6	1.00
205A	2.20	0.05		1.38		642	6	1.00
206	2.20	0.05	0.41	0.96	0.21	1094	6	1.00
207	2.25	0.00	1.00	0.43	0.50	850	6	1.00
208	2.25	0.00	1.00	0.43	0.50	620	6	1.00
209	1.85	0.40	1.00	0.04	0.50	350	6	1.00
210	1.60	0.01	1.00		0.50	250	6	1.00
211	1.60	0.01	1.00		0.50	120	6	0.66
212	0.62	0.00	0.93		0.50	120	6	1.00
213	0.62	0.00	0.93		0.50	40	6	0.46
214	0.99	0.01	0.07			120	6	1.00
215	0.99	0.01	0.07			40	6	1.00

2.1.2. Bunsen and HI_x Sections.

The flowsheets previously developed by General Atomics (Brown L. C., 2003) for both the Bunsen and HI_x sections have been assumed.

Accordingly, the exothermic Bunsen reaction takes place mostly in a heat exchanger reactor operating at 7 bar and 105-120°C. The feed from the HI_x and sulfuric sections is characterized by an excess of water and iodine in order to allow acceptable reaction rates and easy separation of products (H_2SO_4 and HI) into two liquid phases (Sakurai M., 2000): an upper phase essentially constituted by diluted H_2SO_4 and a heavier phase containing HI, water and excess I_2. A gas phase composed of oxygen and residual SO_2 is separated from the two liquid phases in a downstream vessel. H_2SO_4 and HI generation is also achieved in a secondary (boost) reactor and in scrubber towers used for oxygen separation (Brown L. C., 2003).

For what concern the HI_x section a feasible flowsheet has been initially developed by Roth and Knoche (Roth M., 1989) and, after, employed by General Atomics (Brown L. C., 2003). It basically consists of a reactive distillation packed column, operating at 22 bar, fed with the $HI/H_2O/I_2$ stream coming from the Bunsen section, that's pressurized and preheated up to 262°C. The product from the bottom, essentially iodine (I_2) at 310°C, and a $HI/H_2O/I_2$ side stream (at 251°C) are cooled down and recycled to the Bunsen section. Whilst I_2 is concentrated in the bottom section of the column, HI dissociation takes place in the upper part so that the top stream is a gas mixture constituted mostly by H_2 with HI traces that can be quantitatively removed at room temperature in a downstream water scrubber. The hydrogen yield is extremely low if compared with the column hold up (1 mol of H_2 is produced for about 126 mol feed): this is the major drawback of this configuration, that will determine the high costs of the HI_x section (as discussed in the next paragraph) and, consequently, theoretical and experimental research work headed for new solutions is encouraged.

2.1.3. Methanol Section

In a typical plant methanol production is carried out in two steps. The first step is the conversion of the feedstock natural gas into a synthesis gas stream consisting of CO, CO_2, H_2O an H_2. This is usually accomplished by the catalytic steam reforming of the feed gas; partial oxidation is another possible route. The second step is the catalytic synthesis of methanol from synthesis gas. Both these steps can be carried out in a number of ways (Jean P. L., 2001).

In the case here presented the availability of hydrogen allows us to use flue gas delivered by methane combustion as a feedstock for methanol synthesis, according to the following reactions:

$CO + 2H_2 \rightarrow CH_3OH$

$CO_2 + 3H_2 \rightarrow CH_3OH + H_2O$

$CO + H_2O \rightarrow CO_2 + H_2$

The overall section can be divided into three main units, characterized by different operative pressures: methane combustion (1 atm), methanol synthesis (250 atm), and methanol purification (2-10 atm). The flowsheet developed for this section is shown in figure 4 while mass balance and composition of streams are reported in table 3 (with reference to 1 mol SO_2 production).

Regarding the methane combustion, it was assumed to perform a methane partial oxidation using pure oxygen from the Bunsen section.

The choice of methane partial oxidation instead of total oxidation comes from the optimization of overall thermal efficiency. Actually, according to the result of a preliminary analysis, it has been found that methanol synthesis from CO is thermodynamically more convenient than from CO_2: firstly CO_2 is a species more stable than CO and consequently it achieves a lower equilibrium conversion; furthermore CO_2 reaction into methanol consumes a higher amount of hydrogen (3 mol vs. 2 mol) per mole of methanol produced. Therefore, it has been chosen to operate the furnace with a sub-stoichiometric amount of oxygen, obtaining a flue gas mixture essentially consisting of CO with a few amounts (ca. 11 %mol.) of CO_2.

Under the selected operative conditions (320°C, 250 atm) and with 540% of excess hydrogen recycled in reactor R301 (figure 4), the overall conversion of carbon oxides is relatively high, more than 85% for CO and about 70% for CO_2. Outlet stream 309 consists of methanol, steam, carbon oxides and a small amount of methane. After a cooling step (exchanger E304), crude methanol is separated from gases (in gas-liquid separator S302), and sent to the downstream distillation units (columns C301 and C302) operating at pressures far lower than that of reaction section, in which the gaseous and liquid byproducts are separated and high-purity methanol is obtained with molar fraction up to 0.96.

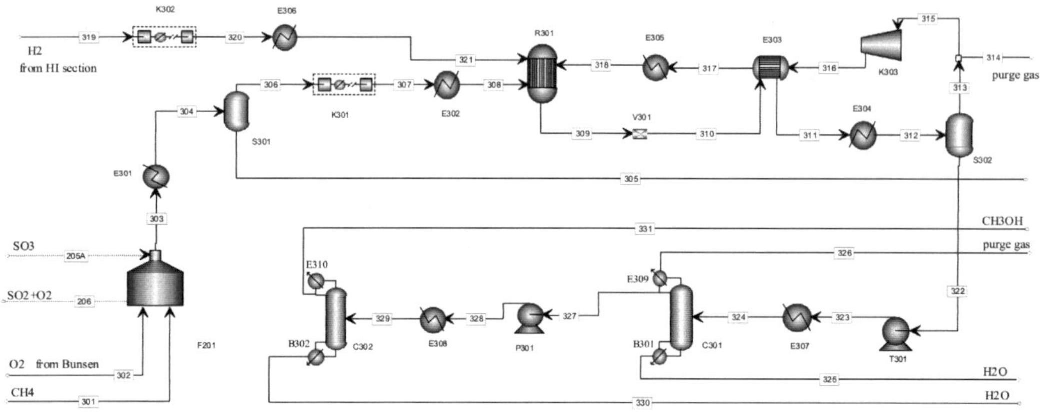

Figure 4. Flowsheet for the methanol plant section.

Table 3. Material balance of methanol synthesis section (1 mol SO_2 production).

Stream	Mole flow (mol/mol$_{SO2}$)							T (°C)	P (atm)	vapour fraction
	H_2O	CH_3OH	CO	CO_2	CH_4	H_2	O_2			
301					0.27			25	1	1.00
302							0.42	25	1	1.00
303	0.53		0.23	0.04	(< 0.01)			120	1	1.00
304	0.53		0.23	0.04	"			30	1	0.35
305	0.52							30	1	0.00
306	0.01		0.23	0.04	(< 0.01)			30	1	1.00
307	0.01		0.23	0.04	"			150	250	1.00
308	0.01		0.23	0.04	"			320	250	1.00
309	0.05	0.27	0.03	0.02	"	3.08		320	250	1.00
310	0.05	0.27	0.03	0.02	"	3.08		320	245	1.00
311	0.05	0.27	0.03	0.02	"	3.08		141	245	0.95
312	0.05	0.27	0.03	0.02	"	3.08		30	245	0.91
313	(< 0.01)	(< 0.01)	0.03	0.01	"	3.07		30	245	1.00
314	"	"	(< 0.01)	(< 0.01)	"	(< 0.01)	(< 0.01)	30	245	1.00
315	"	"	0.03	0.01		3.07		30	245	1.00
316	"	"	0.03	0.01		3.07		33	250	1.00
317	"	"	0.03	0.01		3.07		302	250	1.00
318	"	"	0.03	0.01		3.07		320	250	1.00
319						0.57		30	1	1.00
320						0.57		150	250	1.00
321						0.57		320	250	1.00
322	0.05	0.27		(< 0.01)		(< 0.01)		30	245	0.00
323	0.05	0.27		"		"		32	2	0.00
324	0.05	0.27		"		"		75	2	0.03
325	0.01			"				124	2	0.00
326		(< 0.01)				(< 0.01)		40	2	1.00
327	0.04	0.26						40	2	0.00
328	0.04	0.26						41	10	0.00
329	0.04	0.26						140	10	0.82
330	0.02							181	10	0.00
331	0.01	0.26						137	10	0.00

2.1.4. Energy Consumption and Efficiency

According to the flowsheet developed by General Atomics (Brown L. C., 2003) the heat demand for the HI_x section is about 245.8 kJ/mol$_{SO2}$. Since this energy demand is required at medium-low temperature (<310°C), it can be satisfied exclusively by the solar heat source (with Dowtherm HT and molten salts as energy vectors).

On the other hand the heat demand for the Bunsen section can be considered zero. The heat generated in the exothermic reactors is not suitable for recovery owing to the low operating temperatures (less than 120°C) and, therefore, it can be used for thermal recovery only within the Bunsen section itself.

The heat demand for the sulfuric acid concentration and dissociation subsections are 85.1 and 345.3 kJ/mol$_{SO2}$, respectively; the latter includes the contribution provided by the methane furnace that is 141.9 kJ/mol$_{SO2}$.

Therefore about 79% of the total heat load for the chemical process (676.2 kJ/mol$_{SO2}$) can be supplied by the solar plant.

The whole energy needed for pumping in both HI_x and Bunsen sections is assumed to be 30.0 kJ/mol$_{SO2}$ (Brown L. C., 2003), while the estimated values for the sulfuric and methanol sections are 26.1 and 23.8 kJ/mol$_{SO2}$ respectively: thus, the total energy for electrical power supply is 79.9 kJ/mol$_{SO2}$.

As a result the estimated energetic efficiency, defined as the ratio between energy output in terms of high heat values (HHV) of products (H_2 and methanol, 285.9 and 726.1 kJ/mol respectively) and total energy consumption, is 36.5%:

$$\eta = \frac{X_{CH3OH} \cdot HHV_{CH3OH} + X_{H2} \cdot HHV_{H2}}{Q_{CH4} + Q_{sun} + Q_{el}} = 0.365$$

where X_{H2}, X_{CH3OH} and X_{CH4} are the amounts of H_2, methanol and methane involved (i.e. 0.43, 0.26 and 0.27 mol/mol$_{SO2}$), respectively, while Q_{CH4} (= $X_{CH4} \cdot HHV_{CH4}$), Q_{sun} and Q_{el} are the energy supplied by methane, solar plant and electrical duty (in kJ/mol$_{SO2}$), respectively. Although in the methane furnace a partial oxidation of the fuel gas is performed, the heat value HHV_{CH4} related to its oxidation to CO_2 has been considered (HHV_{CH4} = 890.8 kJ/mol), in order to take into account the intrinsic potential energy associated to the methane consumed.

2.2. Solar Plant Facility

The solar concentrator plant, designed by ENEA, is based on a technologically adapted version of the linear parabolic geometry of SEGS (Solar Electric Generating System), successfully operated at Kramer Junction, California, for about twenty years. The innovations introduced by ENEA mainly concern the use of a non-flammable heat transfer fluid with low environmental impact together with the utilization of a thermal storage device designed to provide thermal power at a constant rate, regardless of variations in solar power availability due to solar irradiation variableness and night/day cycles. A general scheme is shown in figure 5.

Regarding the heat transfer fluid, a molten salts mixture consisting of 60% sodium nitrate and 40% potassium nitrate has been chosen as an energy carrier operating in the temperature range 290-550°C.

The linear parabolic mirrors concentrate the direct sunlight radiation on a linear receiver pipe absorbing radiant energy and converting it to high temperature sensible heat of the molten salts which are collected and sent to a hot tank for heat storage (figure 5). On demand the hot salts are delivered to the chemical process and cooled down to lower temperature levels. After cooling, molten salts are sent to the cold tank and successively pumped to the parabolic troughs for restarting the heat collection loop. The relevant temperature drop (about 260°C) between the hot and cold tanks provides a significant storage capacity in terms of thermal energy (about 1 kWh for 5 liters of molten salts). Research finalized to further improvements of SEGS technology is currently in progress with the development of new parabolic troughs, with the general aim of cost reduction and material improvement.

The receiver pipe (figure 5), located on the focal line of the mirrors, is composed of two concentric cylinders separated by a vacuum gap for thermal insulation: the external glass cylinder represents a protective casing, while the internal steel cylinder absorbs solar energy by means of the heat transfer fluid. A spectrally selective coating compound for receiver tube, developed and specifically designed by ENEA to maximize the efficiency at higher temperatures (550°C), covers the steel pipe and guarantees maximum absorption of solar radiation and reduction of infra-red emissions from hot tubes (Rubbia C., 2005).

Such a solar collection system is easily adaptable to the features of a large number of potential sites in high level solar radiation areas and theoretically permits to provide thermal energy continuously except for protracted cloudy periods (> 2-3 days).

The design of a solar plant involves the evaluation of the active surface of the mirrors (A) and the thermal storage capacity (T), which are functions of the solar radiation level of the site, the process heat requirements, the solar energy collection efficiency and the chemical process operative conditions. Regarding the latter point there are two main alternative routes: to collect all the solar radiation available over the year and, as a consequence of solar irregularities and seasonal fluctuations, to produce hydrogen at a variable rate, or to keep constant the energy delivered to the process varying the contribution of a fossil source and cutting-off solar radiation peaks in the summer.

The second solution has been chosen in order to guarantee a constant hydrogen output and to ease the process control and operation, as the chemical plant does not tolerate large deviations of the process parameters from the designed working conditions.

Furthermore during winter periods, when solar radiation is low, solar plant is shut-down for maintenance operations. This strategy permits to reduce thermal storage capacity and fossil source utilization.

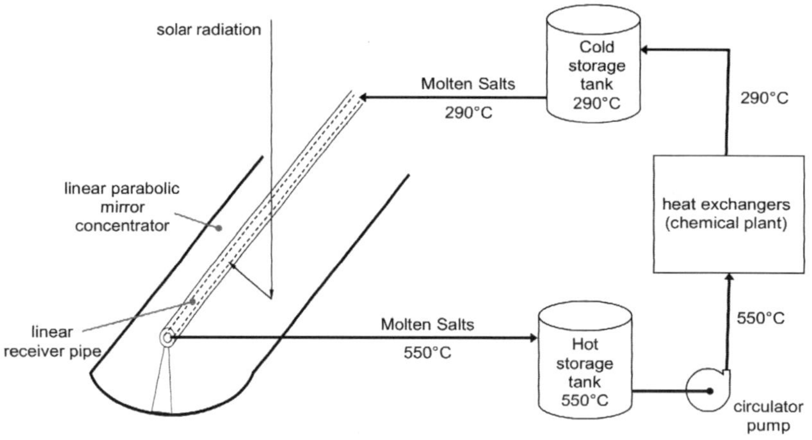

Figure 5. Scheme of the solar parabolic trough concentrator with thermal storage and molten salts loc

3. Plant Sizing and Costing

The economical process evaluation has been done considering an industrial mid-large scale plant with hydrogen capacity of about 86,400 kg/day, coupled to a solar plant with an average thermal power of about 267.1 MW_{th}. Thus the hydrogen output is 500 mol/sec (1 kg/sec), partially converted to about 130 mol/sec of methanol.

For cost estimation a plant life of 20 years has been assumed, with 8% discount rate and 2% rate of inflation.

Results are summarized in table 4.

3.1 Chemical Plant

The single units of the sulfuric section of the chemical plant have been designed and, after, fixed capital costs for the single units have been estimated. The fixed capital investment (FCI) for the other sections of the chemical plant was approximately evaluated too.

In the HI_x section the cost of reactive distillation column and the heat exchangers represents respectively 38% and 44% of the overall section cost, while, for the Bunsen section, the most relevant costs are related to packed columns for HI and SO_2 adsorption and the expansion turbines, which constitute about 36% and 33% of the overall Bunsen section cost.

Also the additional costs related to equipment transportation, installation, piping, instrumentation, control devices, buildings, service facilities, legal expenses and so on, have been roughly estimated and the results are summarized in table 4.

3.2 Solar Plant

In order to optimize the efficiency of the solar concentration system, the plant is supposed to operate continuously for 300 days/year in the period with the maximum insolation. In fact, the Italian site of Priolo (Sicily) has been chosen as reference for yearly solar radiation data and, as the Italian winter solar radiation is low due to long cloudy periods, the programmed plant shut-down period for maintenance is assumed to last about two months (65 days). Particularly, the supposed operative period is supposed to last from 20th January to 15th November and the applied insolation data are referred to 2003 (annual Direct Normal Irradiation (= DNI) = 1,748 kWh/m^2/y).

Solar plant sizing has been done taking into account the following considerations:

- the heat requirement for the process at medium temperature (< 550°C) is about 534.3 kJ/mol_{SO2} and, therefore, the power is 267.1 MW_{th}, as described above;
- the average solar-to-thermal efficiency is approximately 0.655, considering the parabolic troughs optic efficiency and the receiver heat absorbing efficiency for the solar radiation data of reference site.

Furthermore, an annual maximum charge for methane integration of 15% (relative to total solar plant power) has been considered as this value, added to the methane supplied for the sulfuric acid decomposition furnace, leads to carbon dioxide emission sensibly lower than that resulting from a conventional methane steam reforming for hydrogen production.

Given that the solar field active area (A), heat storage capacity (T) and solar plant thermal power are parameters strictly connected, it was determined that, for 15% methane integration, the solar plant must be designed with a heat storage capacity of 9 or 7 GWh if coupled to a solar field active area of 2.5 or 3.0 km^2 respectively. Since the fixed capital investment for the solar field is the largest single cost item, the solution with the smaller solar active area has been chosen: particularly, a heat storage capability of 16 GWh with 2.5 km^2 of solar active area has been assumed, that leads to 12% of methane integration (233 GWh, corresponding to about 64 kJ/mol_{SO2}); under this configuration, and considering the heat demand at high temperature, the total heat supplied by methane is about 30.5% of the overall thermal energy required.

Regarding the aforementioned energy losses related to summer radiation peaks, which are not storable because of the limited capacity of accumulation system, they correspond to a relevant fraction of the total heat collected (494 GWh, i.e. about one-fifth of the total amount). However these energy losses are inevitable, as the solar plant capacity cannot be sized for the highest value of summer radiation, which would imply a strong dependency on fossil source during winter.

In order to overcome this economical drawback, an integrated power plant could be conceived to recover the energy surplus available in summer periods. Nevertheless here we did not consider this latter solution, assuming to lose all the non-storable heat.

The method applied for cost estimation is based on the actual cost calculation of the following items:

- investment cost (FCI): solar plant, field, storage tanks, nitrate salts, methane furnace for energy integration;
- operative cost (Co): maintenance, control, electricity;
- methane cost for energy integration (Cg).

Regarding the solar plant, equipment costs are taken from a previous analysis by ENEA for the Archimede project concerning the construction of a power plant coupled to a solar facility in Sicily.

The estimated cost of solar trough plant per unit area is about 266 $\$/m^2$ including the parabolic troughs, collector pipes, control system and installation work. Taking into account the solar field active area (A = 2.5 km^2), the solar plant cost is approximately 665 M$.

The expense for the field is hardly predictable, as it strictly depends on the geographical location. Assuming an average price of 0.266 $\$/m^2$ and considering that the trough field can be as large as twice the active area of collection (A), the heliostat field might cost about 1.3 M$.

The salts unit cost is assumed to be 0.76 $/kg and, as the amount of nitrates needed for plant operation is about $1.76*10^8$ kg, the overall salts cost is 134 M$.

Concerning storage tanks, it has been calculated that several equipments are necessary in order to keep their size under security threshold for structural resistance. Assuming for the tanks a cost per wall unit area of 1200 $\$/m^2$, and assuming to use 22 tanks with diameter and height respectively of 30 m and 12 m, the total tanks cost is about 66.8 M$.

In addition, the cost of a methane furnace delivering a power of 267.1 MW_{th} has been assessed at 130 M$.

Therefore, it has been calculated that the overall investment cost (FCI) is approximately 997 M$.

The annual operative cost for maintenance, control and electricity, is expected to be a percentage of total investment rate (2% of FCI).

The cost for methane integration is mostly influenced by yearly solar radiation and methane cost fluctuations. As mentioned above, assuming the solar radiation of 2003 as reference, the methane integration here represents 12% of the energy demand at medium temperature. Methane cost depends on transportation expenses, and here the specific value of 0.267 $\$/m^3$ has been assumed.

Therefore the annual cost of the solar plant is approximately 114 M$ (table 4), where the higher single cost item is the solar field, which represents about 2/3 of the total plant value.

It is worth considering that the estimated cost here reported represents an upper limit, as the technology development and the large-scale production of solar equipments might reduce the plant expense.

Moreover the possibility to take advantage of Government finances for energy production at "zero-CO_2" emission can be considered: these consist into bank loans with the discount rate balancing the inflation (i.e. both at 2%) for the initial investment expenses, or tax cuts for the hydrogen produced, or partial free funding. In the first case (zero interest rates) the actual cost for the solar plant could be lowered significantly, as reported in table 4.

Table 4. Estimated costs for hydrogen and methanol production plant with S-I thermochemical cycle powered by hybrid solar/fossil sources. (Plant life: 20 years; discount rate: 8%, inflation: 2%).
* value obtained assuming 2% discount rate (Government finances) between brackets.
[a] referred to the site of Priolo (Italy) from data acquired in 2003. [b] (Wade N.M., 2001)

Solar Plant	
Power	267.1 MW_{th}
Annual Direct Normal Irradiation (DNI) [a]	1,748 $kWh/m^2/y$
Annual Irradiation on Collectors [a]	1,412 $kWh/m^2/y$
Irradiation on Collectors during operative period (300 days: jan 20th to nov 15th) [a]	1,316 $kWh/m^2/y$
Collectors active area	2.5 km^2
Effective Solar Field area	5.0 km^2
Heat Storage Capacity	16 GWh
Fixed Capital Investment (FCI)	997.1 M$
Annual Capital Cost *	86,886 k$/y (49,828 k$/y)
Operative Costs (Co)	19,931 k$/y
Methane Integration Energy	233 GWh/y
Methane Cost (Cg)(0.27 $/m^3 assumed for methane unit cost)	7,022 k$/y
Total Solar Plant Annual Cost *	113,839 k$/y (76,782 k$/y)
Chemical Plant	
S-I Plant	
Annual Hydrogen production	25,920 tons/y = 1,029,240 MWh/y
Fixed Capital Investment (FCI)	483.5 M$
Annual Capital Cost *	42,151 k$/y (24,173 k$/y)
Operative Costs	9,669 k$/y
Methane consumed	78,431,069 m^3
Methane Cost (0.27 $/m^3 assumed for methane unit cost)	20,863 k$/y
Electric Energy consumed	202 GWh/y
Electric Energy Cost (0.065 $/kWh assumed for electric energy unit cost)	13,139 k$/y
Water consumed	233,000 tons/y
Water Cost (0.98 $/m3 assumed for demineralized water unit cost [b])	228 k$/y
Total S-I Plant Annual Cost *	86,050 k$/y (68,072 k$/y)
Hydrogen Cost *	0.194 $/kWh (0.141 $/kWh)
Methanol Plant	
Annual Methanol production	107,962 tons/y = 679,630 MWh/y
Fixed Capital Investment (FCI)	90.1 M$
Annual Capital Cost *	7,855 k$/y (4,505 k$/y)
Operative Costs	1,802 k$/y
Electric Energy consumed	86 GWh/y
Electric Energy Cost (0.065 $/kWh assumed for electric energy unit cost)	5,558 k$/y
Hydrogen consumed	7387 Mmol/y = 587 GWh/y
Hydrogen consumed Cost * (0.194 $/kWh above determined for hydrogen unit cost)	113,937 k$/y (82,567 k$/y)
Total Methanol Plant Annual Cost *	129,152 k$/y (94,432 k$/y)
Methanol Cost *	0.190 $/kWh (0.139 $/kWh)

3.3 Process Assessment

As shown in table 4, the specific cost of hydrogen is 0.19 \$/kWh and can be sensibly reduced to 0.14 \$/kWh if some Government finances are allowed.

These prices are within the range of those obtained with other water splitting processes coupled with renewable energy sources like the solar-powered Zn/ZnO thermochemical cycle (Steinfeld A., 2002) and far lower than the cost of hydrogen from water electrolysis using solar-generated electricity with photovoltaic systems, estimated to be approximately 0.61 \$/kWh (0.38\$/kWh with Government finances) considering a site with the same annual insolation.

On the other hand, the specific cost of hydrogen obtained with such S-I process configuration is definitely much higher than those attained with conventional natural gas steam reforming, which are within the 0.01-0.02 \$/kWh range (Lipman T.E., 2004) although without utilization of renewable sources.

Actually, using the S-I thermochemical process here described (powered by mixed solar/fossil source), the demand for fossil fuels and, therefore the CO_2 emission, can be reduced up to 40% if compared to conventional steam reforming processes. In fact, as resulting from both energy and material balances, the ratio between H_2 produced and CH_4 consumed is 0.42 mol/mol for a conventional steam reforming, 0.25 mol/mol for solar steam reforming ($CH_4 + 2H_2O \rightarrow CO_2 + 4H_2$) (Moller S.), and 0.23 mol/mol for the process here presented. This latter value has been determined considering only the S-I plant (figure 1), with 1.00 mol H_2 output, 0.16 mol CH_4 consumed for heat supply at high temperature (with complete combustion to CO_2), corresponding to 141.9 kJ/mol_{H2}, and 0.07 mol CH_4 for solar plant energy integration, corresponding to 64.1 kJ/mol_{H2}. Note that, in order to properly compare the S-I process here described with steam reforming processes, the previous calculation does not consider the integrated production of methanol, conceived to add value to the carbon oxides and to convert hydrogen into a more suitable energy vector.

Methane consumption can be reduced by replacing the fossil furnace with a solar furnace coupled with a thermal energy storage system suitable for the high temperatures (>550 °C) required for sulfuric acid decomposition, such as a packed bed with ceramic pebbles and a hot air or inert gas circulation system, as described elsewhere for the UT-3 thermochemical cycle (Sakurai M., 1996). Otherwise this process can be much more competitive if the furnace will be replaced with a RDFs (Refused Derived Fuels) combustor, thus recovering the energy from solid wastes, or with a biomass furnace with "neutral" CO_2 emission.

Moreover, since the expenses for both the solar plant and the methane required for integration significantly decrease at higher irradiation intensities, the cost of hydrogen can be considerably reduced if, in place of South Italy, a more insolated region is considered for plant location, like some desert sites in North Africa where DNI = ca. 2,900 kWh/m^2/y (ENEA, 2001).

Finally it should be observed that the cost of methanol is extremely close to the cost of hydrogen (table 4) owing to the lower fixed and operative costs for the methanol production plant, if compared to the costs of the other plant sections (solar and S-I plants), so that the major expense is due to hydrogen consumption.

4. Conclusions

A new hybrid fossil/renewable process for hydrogen generation from water with the S-I thermochemical cycle has been studied. The general scheme previously developed by General Atomic (conceived to be coupled to a high-temperature gas-cooled nuclear power reactor) (Brown L. C., 2003) was assumed for the Bunsen and HI_x sections of the chemical plant, whilst a new flowsheet for the sulfuric acid concentration/decomposition section was adopted. In this case a solar plant facility integrated with a methane furnace (used for the higher temperature services) was employed as heat supplier. Both thermochemical process and solar plant have been optimized, sized and cost evaluated. An additional section for methanol synthesis where the carbon oxides produced from methane combustion are reduced to methanol was designed too in order to partially convert hydrogen into a more easily storable and transportable chemical fuel.

By this route about 70% of the overall thermal energy required by the chemical process can be supplied by the concentrated solar energy, the remaining coming from the fossil source.

The results indicate that this process leads to a hydrogen specific cost (0.14-0.19 $/kWh) the same order of the hydrogen produced by means of other water-splitting processes coupled to renewable energy sources. Moreover, if compared to natural gas steam reforming, a reduction of methane consumption (and therefore carbon oxides emissions) for equivalent hydrogen yield is announced.

Furthermore a more suitable plant location in more insolated regions than the proposed site (South Italy) or a different thermal energy source at the higher temperatures (> 550°C) would significantly reduce the abovementioned hydrogen cost and CO_2 emissions.

References

Brown L.C., Besenbruch G.E., Lentsch R.D., Schultz K.R. Funk J.F., Pickard P.S., Marshall A.C., Showalter S.K. “High Efficiency Generation of Hydrogen Fuels Using Nuclear Power.” General Atomic Project 30047, June 2003.

ENEA (Ente per le Nuove Tecnologie, l'Energia e l'Ambiente, Italy) “Solar Thermal Energy Production: Guidelines and Future Programmes of ENEA”, ENEA/TM/PRES/2001 – June 2001.

Jean P. L., “Methanol synthesis: a short review of technology improvements”. Catalysis Today, 2001, vol. 64, p.3-8.

Kodama T. “High-temperature solar chemistry for converting solar heat to chemical fuels”. Progress in Energy and Combustion Science. 2003, vol.29, p.567-597.

Lipman T.E. “What Will Power the Hydrogen Economy? Present and Future Sources of Hydrogen Energy”, UCD-ITS-RR-04-10, Institute of Transportation Studies – Davis One Shields Ave., University of California, July 2004.

Möller S., Buck R., Tamme R., Epstein M., Liebermann D., Meri M., “Steam reforming of methane rich gas in a solar reactor”, solarpaces.org/20030725EHECO1-84 (available from Solarpaces.org web page).

Roth M., Knoche K.F. “Thermochemical Water Splitting Through Direct HI Decomposition from $H_2O/HI/I_2$ Solutions.” Int. J. Hydrogen Energy, 1989, vol.14, p.545-549.

Rubbia C., Antonaia A., Esposito S. (ENEA) *Surface Coating of the Collector Tube of a Linear Parabolic Solar Concentrator.* Patent n° DE60202142D (2005).

Sakurai M., Bilgen E., Tsutsumi A., Yoshida K. “Solar UT-3 Thermochemical Cycle for Hydrogen Production”. Solar Energy, 1996, vol.57, p.51-58.

Sakurai M., Nakajima H., Onuki K., Shimizu S. "Investigation of 2 liquid phase separation characteristics on the iodine-sulfur thermochemical hydrogen production process". Int. J. Hydrogen Energy, 2000, vol.25, p.605-611.

Steinfeld A. "Solar hydrogen production via a two-step water-splitting thermochemical cycle based on Zn/ZnO redox reactions". Int. J. Hydrogen Energy, 2002, vol.27, p.611-619.

Wade N. M. "Distillation plant development and cost update." Desalination, 2001, vol.136, p.3-12.

Xin Dong, Hong-Bin Zhang et al. "Highly active CNT-promoted Cu-ZnO-Al_2O_3 catalyst for methanol synthesis from H_2/CO/CO_2", Catalysis Letters, 2003, vol. 85, p.237-246.

Membrane Processes for the Sulfur-Iodine Thermochemical Cycle

Frederick F. Stewart, Christopher J. Orme, and Michael G. Jones
Idaho National Laboratory
P.O. Box 1625
Idaho Falls, ID 83415-2208

Abstract

Thermochemical water splitting processes for generating hydrogen have been researched for at least thirty years in which over one-hundred chemical cycles have been proposed, some of which are purely chemical processes and others that contain electrochemical steps. A common theme of all of the proposed cycles is the use of a heat source, potentially as high as 2000 °C. Proposed heat sources include nuclear reactors and solar reflectors. One of the most promising cycles is the Sulfur-Iodine (S-I) process, where aqueous HI is thermochemically decomposed into H_2 and I_2 at approximately 400 °C. Regeneration of HI is accomplished using SO_2 generated from a sulfuric acid decomposition cycle performed at 800-900 °C, thus the cycle actually consists of two concurrent oxidation-reduction loops. As HI is regenerated, co-produced H_2SO_4 must be separated prior to decomposition. Current flowsheets employ a large amount (~83 mol% of the entire mixture) of elemental I_2 to cause the HI and the H_2SO_4 to separate into two phases. Removal of water from this system has the direct result of lowering the required quantity of I_2, thus reducing the amount of material that must be physically moved within an S-I plant.

Efforts at the INL have focused on studying the application of pervaporation to the de-watering of an HI/I_2/water feedstream. Experiments have been conducted using Nafion-117® membranes at temperatures between ambient and 110 °C. At 22 °C, water flux was measured at 430 g/m^{2}h. Experiments were conducted at 50, 70, and 100 °C where fluxes increased to 600 g/m^{2}h, 900 g/m^{2}h, and 1500 g/m^{2}h, respectively. Separation factors were measured at approximately 200-400 and decreased slightly with increasing temperature. At higher concentrations of HI in the feed, fluxes dropped significantly due to the lower water content, however the separation factors remained high (100-400, depending on temperature). Membrane degradation was studied using instrumental means. Differential Scanning Calorimetry (DSC) and Thermogravimetric Analysis (TGA) yielded no indication of degradation. However, tan-delta measurements from Dynamic Mechanical Analysis (DMA) did reveal structural changes within the membrane. These changes were not reflected in the membrane transport behavior.

Introduction

Thermochemical water splitting cycles are methods that have been proposed to generate hydrogen on a commercial scale. In general, hydrogen today is formed from the steam reforming of natural gas. Hydrogen generated using this method is consumed in industrial processes, such as petroleum refining. Interest exists in using hydrogen as an energy carrier for the transportation energy sector and other applications where petroleum is used today.

Formation of hydrogen from thermochemical cycles has been proposed using nuclear reactor supplied thermal energy.[1] The only chemical input is water and the only effluent is oxygen and hydrogen. From an energy input standpoint, heat, in the range of 750 °C to 1000 °C, is provided to drive the reactions, which tend to be endothermic. Thermochemical cycles were principally investigated from the 1960's through the 1980's. However, interest waned with the drop in crude oil prices in the 1980's. Recent energy price increases and U.S. national security issues have renewed interest in hydrogen production. Advantages of nuclear driven hydrogen production include hydrogen's inherent renewable nature, a reliance of the process on U.S. domestic resources, and an absence of CO_2 emissions. Similar advantages can be obtained through the use of high temperature solar concentrators.

In the literature, more than 100 thermochemical cycles have been proposed and studied to some degree.[2] One of the more developed cycles, the sulfur-iodine process (S-I), [3], is comprised of three chemical reactions. Water enters the cycle as a reactant in the Bunsen reaction [4], as shown in Eq 1.

$$SO_2 + I_2 + 2H_2O \rightarrow H_2SO_4 + 2HI \qquad \text{eq.1}$$

In Eq.1, hydriodic acid (HI) is formed from the reduction of I_2 at an approximate temperature of 120 °C. HI is then decomposed to generate H_2 at 400 °C to 500 °C, Eq.2, and the co-generated I_2 is recycled. To re-generate SO_2, H_2SO_4 is decomposed at temperatures as high as 900 °C. Eq 3.

$$2HI \rightarrow H_2 + I_2 \qquad \text{eq.2}$$

$$2H_2SO_4 \rightarrow 2H_2O + 2SO_2 + O_2 \qquad \text{eq.3}$$

Key to the operation of this proposed cycle is the Bunsen reaction. Current flowsheets indicate that 7500 Kg of material must be physically moved within a plant for every Kg of hydrogen generated, and much of this mass is iodine associated with the Bunsen reaction. A large excess of I_2 is added to separate the HI from the H_2SO_4, which causes the product mixture to separate into two phases. Water and H_2SO_4 comprise the lighter phase, while the heavier phase is a mixture of HI, I_2, and water at an approximate ratio of 2:8:10, respectively. To facilitate this phase separation, a dehydration loop employing concentrated phosphoric acid has been proposed to be added to the system to remove water from the HI/I_2 mixture. Direct removal of water using a membrane process would allow for a large reduction in the amount of iodine, thus increasing the efficiency of the cycle by lowering the recycle ratio per Kg of hydrogen generated. However, membrane concentration of HI/I_2 mixtures has not been demonstrated. For these initial studies, we have focused on Nafion-117® membranes in the pervaporation of water from a water/HI/I_2 feed.

Experimental Methods

Nafion-117® membranes were obtained from Aldrich Chemical Co. and used as received. Membrane thicknesses were measured using a Mitutoyo micrometer to be approximately 180 µm. A lower concentration feed was prepared by dilution of 55 % aqueous solution of HI (52 g, Aldrich Chemical Co.) with deionized water (118 g). The higher concentration feed consisted of the 55 % HI as received without dilution. HI concentrations in both the feed and permeate samples were determined by titration against standardized sodium

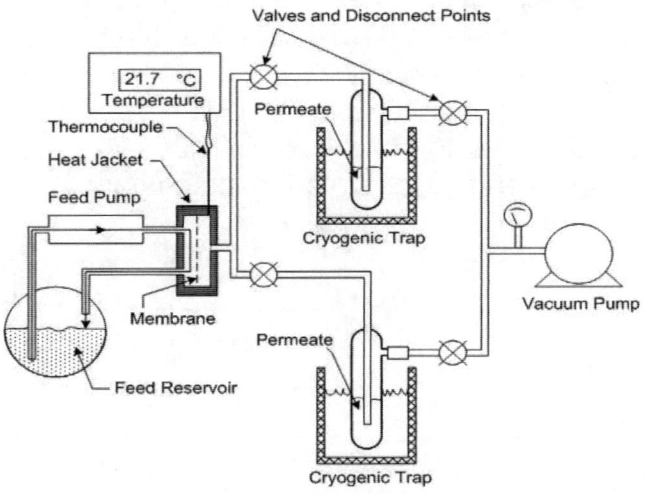

Figure 1. Pervaporation System

hydroxide solutions to a phenolphthalein endpoint. A pervaporation apparatus, see Fig. 1, was constructed for this work using a Millipore 47 mm filtration cell modified with an extra port for feed side flow.

Membranes were secured using Viton O-rings and leak-checked against 50 psi nitrogen gas prior to use. All tubing and connectors were made of PTFE. Pumping of the feed was accomplished using a PTFE peristaltic pump obtained from Cole-Parmer Inc. Temperatures were measured at the cell using a properly calibrated thermocouple. The temperature of the cell was controlled with heat tape. Feed flow rates were sub-turbulent at approximately 50 ml per minute. Permeate side pressure was maintained at 100 mTorr using a mechanical vacuum pump. All permeates were trapped cryogenically and measured gravimetrically to determine the transmembrane flux. In this work, all feeds were treated as batches where the pervaporation system was operated for the amount of time needed to reduce the feed volume to an amount that could no longer be effectively pumped.

Results and Discussion

Pervaporation experiments using Nafion-117® membranes were conducted at temperatures ranging from 22 °C to 100 °C on feed fluids that ranged from approximately 18 to 46 % HI in water (lower concentration). Initial membrane flux, the amount of material that the membrane passes, was found to increase with increasing temperature. At 22 °C, flux was measured at 0.43 Kg/m^{2}h. Experiments were conducted at 50 °C, 70 °C, and 100 °C where fluxes increased to 600 g/m^{2}h, 900 g/m^{2}h, and 1500 g/m^{2}h, respectively. An Arrhenius plot of this data, Fig. 2, revealed the activation energy of flux (E_j) of 14.8 KJ/mol. Experiments conducted at higher feed concentrations, between 50 and 70 wt. % HI, revealed a slightly higher E_j, which is reasonable considering the lower water content of the feed.

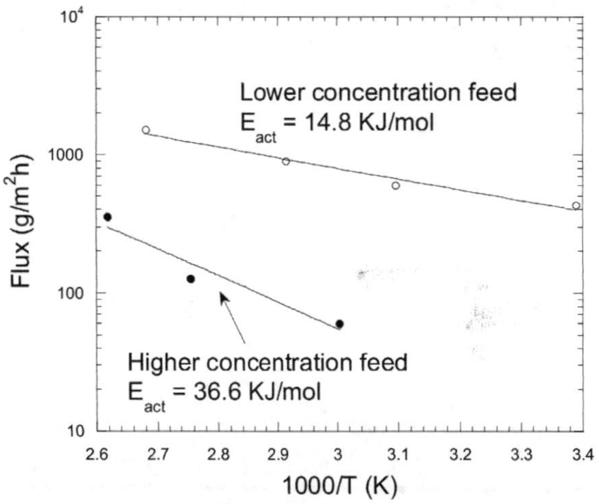

Figure 2. Arrhenius plot of membrane performance with higher and lower concentration feed.

The significant aspects of membrane performance are transmembrane flux and the separation factor. All experiments were conducted in a batch configuration such that the concentration effect could be effectively measured and correlated to both flux rate and separation performance. The results of the pervaporation experiments are expressed in terms of the feed concentration. As expected, the fluxes decrease with increasing concentration of HI. For the lower HI concentration experiments, Fig. 3, HI was effectively concentrated from approximately 18 % to 46 %. At the relatively higher temperatures of 70 °C and 100 °C, fluxes were increased resulting in more effective concentration of the acid, where at lower temperatures, the concentration effect required more time due to lower fluxes. Thus, as expected, flux is directly correlated to temperature and inversely correlated to the content of water in the feed.

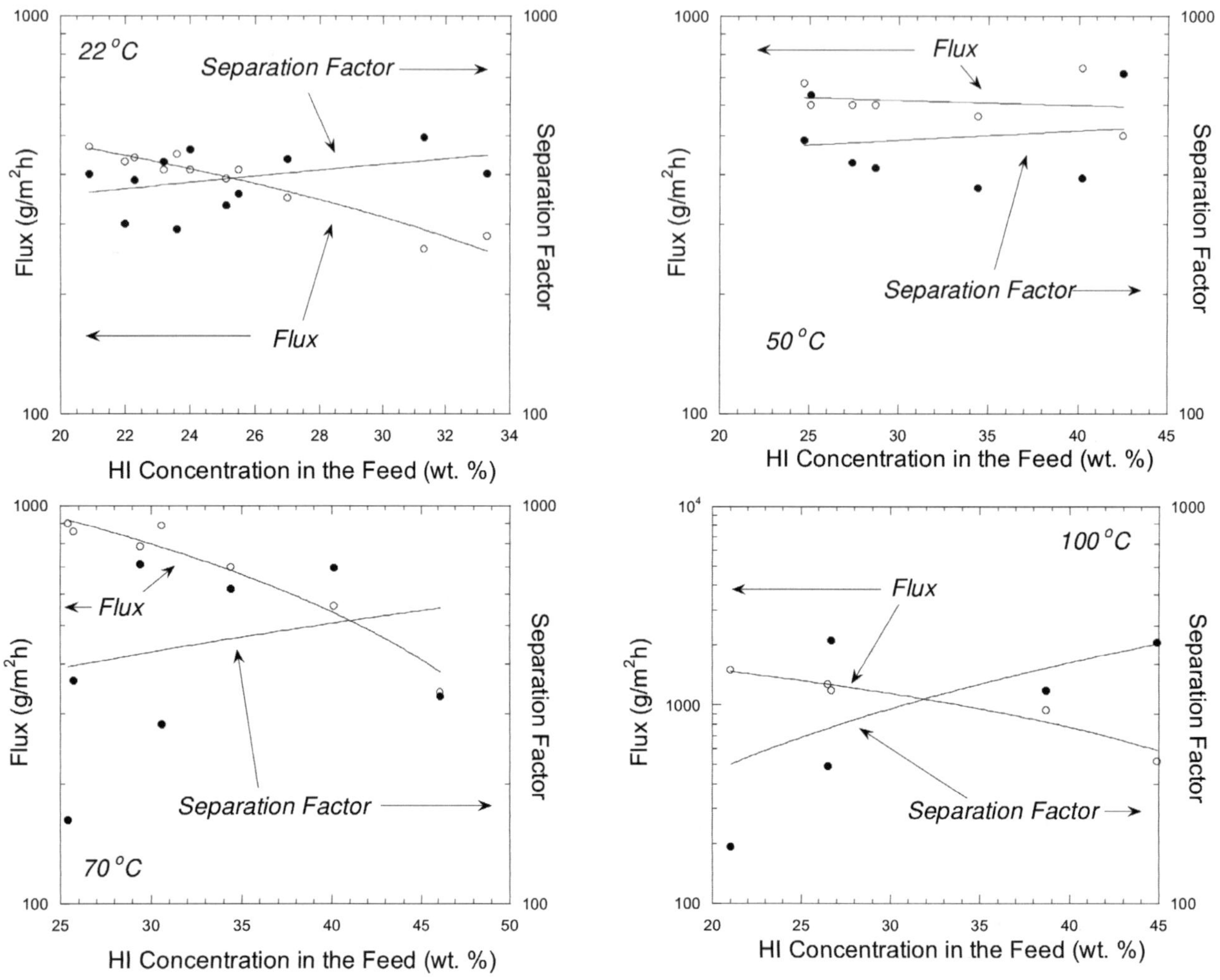

Figure 3. Membrane fluxes and separation factors for Nafion-117® membranes at 22 °C, 50 °C, 70 °C, and 100 °C.

Separation factors universally were observed to increase with increasing concentration of HI in the feeds. As the water becomes depleted upstream of the membrane, increased permeation of HI and I_2 is not observed. Separation factors were also observed to decrease slightly with increasing temperature. For example, at 22 °C, the initial separation factor was 301 and decreased to 139 at 100 °C. When considered with the flux rate enhancement with temperature, this result is typical of membrane performance in that a typical trade-off between flux and separation factor is observed. In general, enhancements in permeability often result in a loss of selectivity.

Pervaporation of more concentrated feeds was accomplished at 60 °C, 90 °C, and 109 °C. In these experiments, fluxes were substantially lower initially than the previous data, which is reasonable considering the lower content of water in the feed. Furthermore, the previously noted trend of lower fluxes as the feed became more concentrated was also observed for this set of experiments. However, within this data set, initial fluxes were higher with higher operating temperatures, consistent with previous data. Separation factors also revealed the same trends as the previous experiments with both increasing HI concentration in the rejectate and increasing operating temperature. At 60 °C, initial separation factors ranged between 300 and 400. Increasing the temperature to 90 °C and 109 °C gave separation factors between 200-700 and 200-500, respectively. Again, the typical membrane trade-off between flux (permeability) and separation factor (selectivity) is observed.

Membrane durability has been studied using thermomechanical methods. Differential scanning calorimetry (DSC) and thermogravimetric analysis (TGA) reveal no significant changes to the membrane upon use. Defects that would be detected using these methods would be changes in crystallinity or chain scission processes. Dynamic mechanical analysis (DMA), an inherently more sensitive technique for polymer degradation than either DSC or TGA, indicated some increased stiffening of the membrane consistent with either chemical cross-linking or inclusion of ionic species within the polymer matrix. Most significantly, these mechanical changes were not observed to change the transport behavior of the bulk membranes.

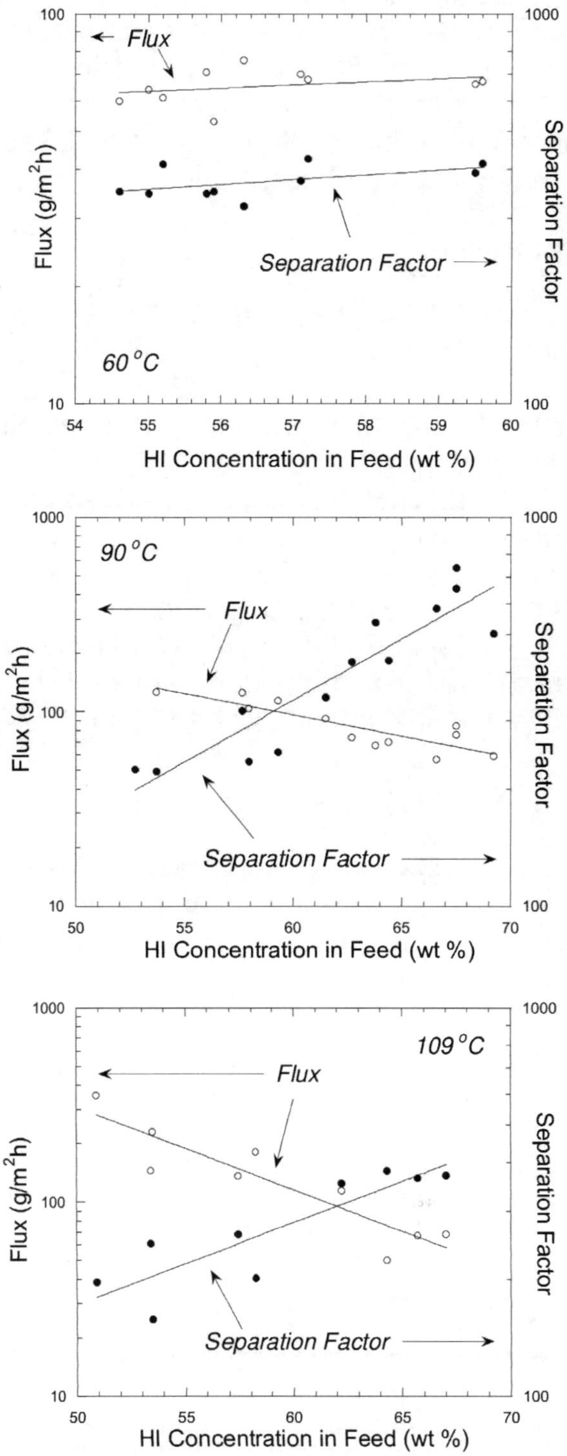

Figure 4. Membrane performance data at 60 °C, 90 °C, and 109 °C.

Conclusion

Nafion-117® membranes have been utilized in pervaporation to separate water from a water/HI/I_2 mixture. High separation factors were observed for all temperatures studied indicating that this membrane material may be appropriate for this application. However the desired temperature range for a membrane system is between 120 °C and 350 °C, and Nafion-117® has yet to be characterized in this temperature range. A trade-off between the water permeability and the selectivity of the membrane has been observed with increasing system temperature. It is expected that this trade-off will continue as the system temperatures are increased further. However, at this time, Nafion-117® appears promising in that little degradation is observed upon usage indicating that the perfluoro-sulfonate nature of the polymer is resilient to acid induced cleavage. Additionally, good water fluxes are observed that are competitive with other pervaporation processes, indicating the potential economic employment of this technology.

References

[1] J. H. Norman, G. E. Besenbruch and D. R. O'Keefe, Thermochemical water-splitting for hydrogen production, General Atomics Report GRI-80/0105, 1981

[2] L. C. Brown, G. E. Besenbruch, R. D. Lentsch, K. R. Schultz, J. F. Funk, P. S. Pickard, A. C. Marshall and S. K. Showalter, High efficiency generation of hydrogen fuels using nuclear power, General Atomics Report GA-A24285, 2003

[3] J. H. Norman, G. E. Besenbruch, D. R. O'Keefe and C. L. Allen, Thermochemical water-splitting cycle, bench-scale investigations, and process engineering, final report for the period February 1977 through December 31, 1981, General Atomics Report GA-A16713, DOE Report DOE/ET/26225-1, 1982.

[4] M. E. Davis and W. L. Conger, An entropy production and efficiency analysis of the Bunsen reaction in the general atomic sulfur-iodine thermochemical hydrogen-production cycle, Int. J. Hydrogen Energy, 5 (1980) 475.

Acknowledgment

This work was supported by the U.S. Department of Energy, Office of Nuclear Energy, Science, and Technology, and the INL Laboratory Directed Research & Development (LDRD) Program under DOE-NE Idaho Operations Office Contract DE-AC07-05ID14517.

75d Construction Material Development in Sulfur-Iodine Thermochemical Water-Splitting Process for Hydrogen Production

Ajit Roy, Benjamin E Russ, Bob Buckingham, Bunsen Wong, Gottfried Besenbruch, Lloyd C Brown

Among the different thermochemical hydrogen processes, the Sulfur-Iodine (SI) water splitting cycle is a prime candidate as it offers a highly efficient process that can be achieved under practical reaction settings. To realize this on an industrial scale, suitable materials of construction that can durably withstand the extremely corrosive environment must be identified. This presentation will outline the construction material requirements and possible candidates for the different process sections within the SI cycle. Ongoing material screening tests have indicated that refractory metals will be the stand out choices for the HI decomposition process but further developmental testing is required. The results from this screening and other simulated environmental mechanical tests, along with a roadmap outlining the materials development effort will also be presented.

HI Decomposition- A comparison of reactive and extractive distillation techniques for the Sulfur-Iodine Process*

Ben Russ, Bob Buckingham, Lloyd Brown, Bunsen Wong and Gottfried Besenbruch
Inertial Fusion Technology Group
General Atomics
3550 General Atomics Court, Bldg. 22T
San Diego, CA 92121

The Sulfur-Iodine (S-I) cycle consists of three sections integrated to produce hydrogen from water in a multi-step process. In Section I, the Bunsen reaction, sulfur dioxide (SO_2) and iodine (I_2) are reacted with water to produce sulfuric acid and hydriodic acid (HI). In Section II the sulfuric acid is decomposed to regenerate SO_2. Section III involves the decomposition of HI to hydrogen and iodine.

There are two alternative designs for the HI decomposition section. The first method, involves the simultaneous concentration and decomposition of HI in a reactive distillation column. The second method for HI concentration and decomposition utilizes H_3PO_4 to first remove the I_2 from the HI_x stream (48.9:8.1:43 mol % for I_2:HI:H_2O), then subsequently break the HI-water azeotrope to allow distillation of pure HI prior to decomposition. This process is referred to as extractive distillation. It was originally developed at General Atomics (GA) in the late 1970s and early 1980s. While the principle components of the extractive distillation process were demonstrated previously, a continuous flow laboratory demonstration has not been previously shown.

The flowsheet for the extractive distillation was developed at GA in 1979. For the laboratory demonstration, an abbreviated flowsheet was developed using AspenPlus which includes the major components. Based on the flowsheet results, an experimental apparatus was designed.

The system was been designed to demonstrate:

1. 100-1000 l/hr H_2 production
2. HI-water extraction using H_3PO_4.
3. HI distillation using H_3PO_4.

The reactive distillation system was also designed to produce 100-1000 liters per hour of hydrogen. In reactive distillation, the separation of HI from I_2 is combined with HI decomposition in one step. This reduces the amount of HI that must be recycled back to the Bunsen reactor from Section III.

The reactive distillation of HI has not been experimentally demonstrated before. The following factors were considered for the design of the reactive distillation apparatus at GA:

- The first distillation column was constructed from glass, as the HI_x solution is highly corrosive. (The Bench-Scale distillation column will be designed from metals that have shown excellent performance in prior corrosion experiments).

- The pressure (40 bar) required to distill HI requires that the glass column be placed in a pressurized environment. This external pressure allows the glass column to operate without large pressure gradients across the glass walls of the apparatus.
- The distillation column must be insulated to ensure adiabatic operation.

General Atomics is developing a laboratory demonstration of both the reactive and extractive distillation systems for HI decomposition as part of a NHI project in conjunction with Sandia National Laboratories and the Commissariat à l'Energie Atomique (CEA) in France. Preliminary data for the two systems will be presented along with a detailed comparison of the two methods.

*Work supported by U.S. Department of Energy under Contract DE-FG06-03RL14520.

75f Applicability of Inorganic Membranes for Improving the Sulfur-Iodine Process for the Production of Hydrogen Using Nuclear Energy

Brian L Bischoff, Charles Forsberg, Dane F Wilson, Lee D Trowbridge, Louis K Mansur

The Sulfur-Iodine (SI) thermochemical process is the leading candidate worldwide for production of hydrogen (H_2) using heat from nuclear energy. This process thermocatalytically cracks water yielding hydrogen and oxygen. The SI process consists of a series of chemical reactions where all the chemicals are recycled in the process except for water. The SI process is efficient, scalable to large sizes, and uses no expensive chemical reagents; however, it has one major disadvantage: high operating temperatures (800 to 1000 C). The high-temperature chemical reaction in this cycle is the thermal decomposition of sulfuric acid. There is a potential to lower the peak temperature by 200+ C if the high-temperature decomposition products of sulfuric acid, O_2, H_2O, and SO_2, can be separated from SO_3 using an inorganic membrane. First generation membranes have been fabricated from materials selected based on literature stability data. Results on the stability of these membranes under simulated operational conditions and data showing the potential to separate the product gases from SO_3 will be presented.

Research was sponsored by the Laboratory Directed Research and Development Program of Oak Ridge National Laboratory and the U.S. Department of Energy's Nuclear Hydrogen Initiative through the Office of Nuclear Science and Technology. The Oak Ridge National Laboratory is managed by UT-Battelle, LLC, for the U.S. Department of Energy under Contract No. DE-AC05-00OR22725.

75g Quantitative Analysis of the Sulfur-Iodine Cycle Through Process Simulation*

Lloyd C Brown, Paul M Mathias

Hydrogen production by thermochemical water splitting, a process that accomplishes the decomposition of water into hydrogen and oxygen, is an environmentally attractive way to produce hydrogen without using fossil fuels. The Sulfur-Iodine (S-I) water-splitting cycle is a promising candidate for thermochemical hydrogen production. The S-I Cycle involves complex phase and chemical equilibria and a highly integrated flowsheet, and while many of the process steps have been studied in the laboratory, an integrated pilot unit operating under prototypic conditions has yet to be built. Hence, process simulation offers a potentially useful way to understand and optimize the SI Cycle. This work has built upon previous efforts in physical-property flowsheet modeling to develop an integrated Aspen Plus simulation of the SI Cycle.

The paper will highlight the many simulation problems that were overcome and demonstrate the value that can be obtained from a robust process simulation. Some of the benefits of the process simulation are: reliable estimates of the energy efficiency of the process; rational improvement of the energy efficiency; and understanding of process issues, like safety.

*Work supported by U.S. Department of Energy under Contract: DE-FG07-03RL-14520.

76a Solar Thermochemical Production of Hydrogen

Aldo Steinfeld

Solar thermochemical processes, which make use of concentrated solar radiation as the energy source of high-temperature process heat, are considered for the production of solar fuels. Two thermochemical routes for solar hydrogen production are examined:

1) H2O-splitting cycle using ZnO/Zn redox reactions.

2) Decarbonization of fossil fuels by thermal cracking, gasification, and reforming reactions.

A 2nd-Law analysis is conducted to determine maximum exergy efficiencies and CO2 mitigation potential. The solar-driven decarbonization process avoids the discharge of pollutants or the contamination of the products, and further upgrades the calorific value of the fossil fuel.

Three prototype solar chemical reactors were developed:

a) a rotating cavity-receiver lined with ZnO-particles was used for the thermal dissociation of ZnO at above 2000 K;

b) a 2-cavity reactor using a graphite absorber was used for the carbothermal reduction of ZnO at 1500 K;

c) a CH4 vortex flow laden with C-particles was used for the thermal cracking of natural gas at 1500K.

The reactor configurations are shown in the adjacent figure. These reactors feature efficient heat transfer by direct irradiation of the chemical reacting systems. Their engineering design and experimental performance are presented.

Scheme of solar reactor configurations for (a) thermal dissociation of ZnO, (b) carbothermic reduction of ZnO; and (c) thermal cracking of CH4.

Stability of Sulfuric Acid Decomposition Catalysts for Thermochemical Water Splitting Cycles

Daniel M. Ginosar, Anne W. Glenn, and Lucia M. Petkovic
Idaho National Laboratory
P.O. Box 1625
Idaho Falls, ID 83415-2208

ABSTRACT

Thermochemical cycles produce hydrogen through a series of chemical reactions where the net result is the production of hydrogen and oxygen from water at much lower temperatures than direct thermal decomposition. The sulfur-based family of thermochemical cycles appears to have significant technical viability for producing hydrogen. The sulfur-based cycles employ a high-temperature catalytic sulfuric acid decomposition reaction step. The reaction produces oxygen and generates SO_2, which is used in other reaction steps of the cycle. The reaction takes place from 750 to 900 °C, or higher, and is facilitated by a heterogeneous catalyst.

The activity and stability of several supported platinum catalysts were explored. Reactions were carried out using a feed of concentrated liquid sulfuric acid (96 wt%) at atmospheric pressure at temperatures between 800 and 850 °C and a weight hour space velocity of 52 g acid/g catalyst/hr. Reactions were run at these high space velocities such that variations in kinetics were not masked by surplus catalyst. Higher surface area Pt/Al_2O_3 and Pt/ZrO_2 catalysts were found to have the highest activity but deactivated rapidly. A low surface area Pt/TiO_2 catalyst was found to have good stability maintaining its activity for over 200 hours of continuous operation.

INTRODUCTION

All of the sulfur-based thermochemical cycles, including the Sulfur-Iodine cycle, the Hybrid Sulfur cycle, and the Sulfur-Bromine Hybrid cycle, incorporate the sulfuric acid decomposition reaction as the high temperature step in their cycles. This reaction decomposes sulfuric acid to oxygen, steam, and sulfur dioxide. The production of oxygen represents half of the water splitting reaction and is a product from the cycle. The generated steam and SO_2 are recycled to the processes and are used in the other reaction steps to complete the water splitting reaction to produce hydrogen.

The sulfuric acid decomposition reaction takes place at temperatures between 750 and 900 °C, or higher, [1] and is catalyzed by a heterogeneous metal oxide or precious metal catalyst. This reaction is the highest temperature unit operation in the thermochemical hydrogen production process and dictates the temperature requirements of the primary energy source. High temperatures for the sulfuric acid reaction are necessary to achieve high levels of chemical conversion, and hence energy efficiency and rapid reaction kinetics.

Catalysts for sulfuric acid decomposition were explored during the 1970s and 1980s [2-7]. Activities of over 30 catalysts have been reported. However, the majority of these studies examined catalyst behavior under ideal conditions, including very dilute acid solutions, excess catalyst, low space velocities and short experimental times [2-6]. Those studies found that there were few highly active catalysts. Most metal oxides had low activity due to the formation of sulfates on the catalysts [3-6] The formation of sulfates was more significant at the lower reaction temperatures and was predicted to be problematic as reaction pressure increased to anticipated operating values [5,6]. In addition, catalysts were found to fail due to formation of volatile acid salts [4], support poisoning [4], and catalyst attrition [7]. Although a few catalysts appeared promising under ideal conditions, primarily platinum supported on porous metal oxides, their long-term stability has not been reported. Catalyst stability is very important since the decomposition reaction environment is extremely harsh where catalysts are exposed to conditions of high-temperatures and aggressive chemicals, including high-temperature steam, oxygen, SO_3, and SO_2. Since the Pt based catalyst were previously reported as the most promising, this effort focused on platinum/metal oxide supported materials and examined their catalytic activity and stability.

EXPERIMENTAL METHODS

A high-temperature continuous flow catalyst test system was used to examine the catalytic reaction in flowing concentrated sulfuric acid at atmospheric pressure. The system, built out of Teflon tubing and glass and employing a high-temperature quartz reactor, pumps concentrated sulfuric acid into a heated zone where the acid is vaporized and decomposed to SO_3 and steam. The vapors flow over a bed of catalyst for the reduction of SO_3 to produce SO_2 and oxygen. Any unreacted SO_3 exiting the reactor recombines with H_2O to form sulfuric acid and is condensed and collected. The remaining gasses bubble through a solution of 1N NaOH to adsorb SO_2 while oxygen is vented to a hood. The system is protected from overpressure via pressure relief valves, one prior to the entrance of the reactor and a second in the product collection section. The concentration of the collected H_2SO_4 was determined by a standard acid-base titration. The concentration of the SO_2 collected in the NaOH solution was determined by a standard iodometric titration.

Initial catalyst testing focused on three supported platinum catalysts, Pt/Al_2O_3, Pt/TiO_2, and Pt/ZrO_2, with platinum loadings of 0.1 wt %. Catalysts were obtained from Johnson Mathey Co. The Al_2O_3 support was in the gamma phase. Tests were performed at 800 and 850 °C at atmospheric pressure. Liquid sulfuric acid at 96 wt% was pumped into the catalytic reactor at 52 g/hr. One gram of catalyst was loaded into the reactor resulting in a weight hour space velocity (WHSV) of 52 hr^{-1}. Catalyst activity was monitored with time on stream (TOS) and catalyst properties were examined *ex situ* following the reaction study.

Nitrogen physisorption measurements were performed on an automated Quantachrome Autosorb-1C system. Fresh and spent samples were outgassed at 673 K for 3 hours before the nitrogen physisorption measurements. Prior to outgassing, the fresh catalyst samples were heated in-situ under flowing helium from room temperature to 1073 K in 2 hours to resemble the initial pretreatment conditions of the samples submitted to reaction. B.E.T. surface areas were calculated in the range of P/Po between 0.05 and 0.35 and total pore volume was determined at P/Po = 0.99.

Diffuse reflectance infrared Fourier transform spectroscopy (DRIFTS) analyses were performed on a Nicolet Magna 750 Fourier transform infrared spectrometer equipped with a commercial Spectratech diffuse reflectance cell. Fresh and spent samples were pretreated *ex situ* under the same protocol as the one utilized for nitrogen physisorption measurements. The samples were transferred to the DRIFTS cell and treated under flowing nitrogen for 1 hour at 383 K. Carbon monoxide was allowed to flow through the catalyst bed kept at 383 K for 5 min and then the weakly adsorbed CO molecules were removed by flowing nitrogen for 30 min. Eight-hundred-scan DRIFTS spectra were collected in the 4000-400 cm^{-1} range at a resolution of 4 cm^{-1}.

RESULTS

Activity and stability results for the three catalysts over 6 h are shown in Figures 1 and 2 for temperatures of 800 and 850 °C, respectively. At 800 °C the highest initial SO_2 yield conversion of about 45 % was obtained using the zirconia supported catalyst, with the alumina and titania supported catalysts providing yields of approximately 30 and 20 %, respectively. At the higher temperature of 850 °C, the highest initial yield was obtained using alumina, followed by zirconia and titania.

At 800 °C the Pt/ZrO_2 catalyst had the highest level of deactivation with yield dropping from 45 to 30 % in about 6 h. The activity of the Pt/Al_2O_3 catalyst also dropped about 10 % over the same time period while the Pt/TiO_2 catalyst was stable. At 850 °C the Pt/Al_2O_3 catalyst demonstrated the largest activity decline of over 15 % while the Pt/ZrO_2 catalyst had a modest decline and the Pt/TiO_2 catalyst appeared to have an increase in activity with time.

Since the Pt/TiO_2 catalyst appeared to be relatively stable, a longer-term stability test was conducted. Results are shown in Figure 3. Yield of SO_2 on the first day on stream averaged 35%. Yield increased to about 42% on days 3 and 4 and then declined to about 34% by the tenth day of testing. Although activity variations were present, the data clearly demonstrates that the Pt/TiO_2 catalyst had significant activity for over 200 hours of continuous operation.

Ex situ catalyst analysis following these reaction studies provides some insight into the deactivation process. Upon removal of the catalyst from the reactor, two visual results were obvious: the Pt/ZrO_2 catalysts run for 6 h at both 800 and 850 °C broke down to produce fines, and; the Pt/TiO_2 catalyst run for over 200 h had changed color from gray to white. The physical degradation of the Pt/ZrO_2 catalyst was also previously observed [7] and could lead to significant equipment plugging problems in production units. The color change on the Pt/TiO_2 catalyst may have resulted from the loss of surface platinum, or the formation of sulfates.

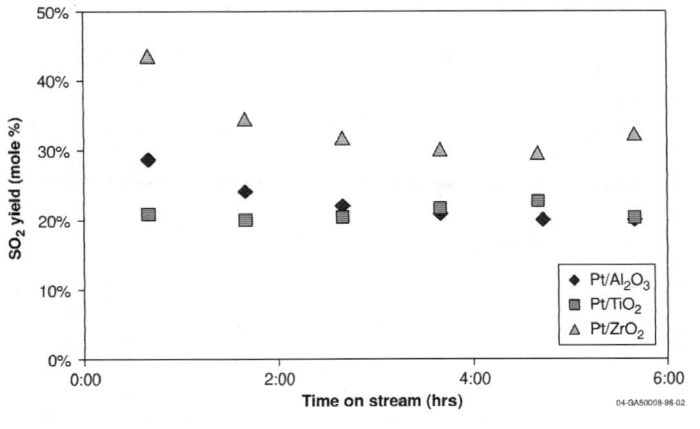

Figure 1. Catalyst activity at 800 °C.

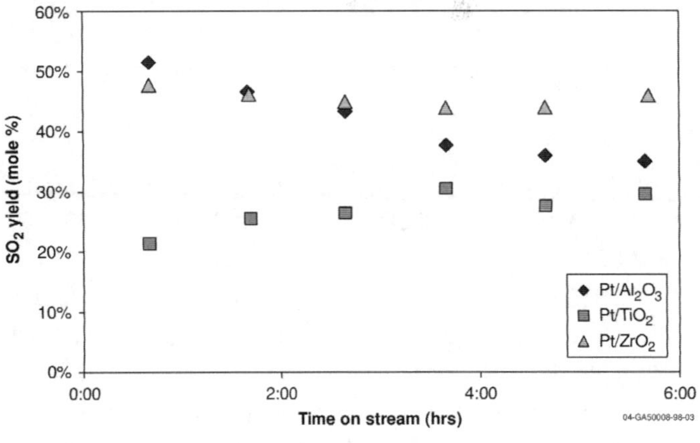

Figure 2. Catalyst activity at 850 °C.

Change in catalyst surface area was analyzed by nitrogen physisorption. Results are shown in Table 1. Surface areas for the fresh catalysts were approximately 180, 50, and 2 m^2/g for the Pt/Al_2O_3, Pt/ZrO_2, and Pt/TiO_2 catalysts, respectively. The low surface area of the Pt/TiO_2 catalyst was probably responsible for the lower catalyst activity. In general, higher surface area catalysts had higher initial reaction rates. The surface area of the Pt/Al_2O_3 and Pt/ZrO_2 catalysts decreased significantly upon exposure to reaction conditions for 6 h. Higher reaction temperature resulted in greater loss in surface area. The changes in surface area for the Pt/TiO_2 catalyst are within the instrument measurement error range and are not considered significant.

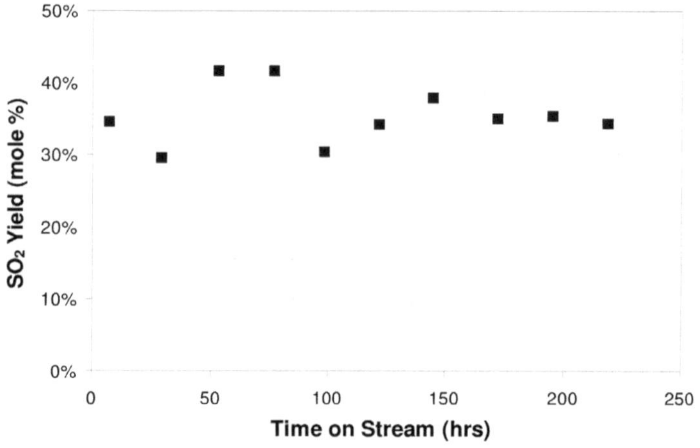

Figure 3. Pt/TiO_2 catalyst stability at 850 °C for over 200 h time on stream.

Table 1. Catalyst surface areas before and after reaction.

Catalyst	Fresh (BET SA, m^2/g)	Reaction at 800°C (6 h) (BET SA, m^2/g)	Reaction at 850°C (6 h) (BET SA, m^2/g)	Reaction at 850°C (>200 h) (BET SA, m^2/g)
Pt/Al_2O_3	179	77	63	—
Pt/ZrO_2	51	25	23	—
Pt/TiO_2	2.0	2.8	2.4	1.8

Analysis by X-ray diffraction (XRD) found a change in support crystalline structure only for the Al_2O_3-based catalyst. Energy dispersive spectroscopy (EDS) used in conjunction with a scanning electron microscope revealed that sulfur compounds were present on the Al_2O_3-based catalyst, but the results were less clear for the other two catalysts.

Catalyst samples were also analyzed by diffuse reflectance infrared Fourier transform spectroscopy (DRIFTS) using carbon monoxide (CO) as a probe molecule (Figure 4). A peak representing CO adsorbed on metal sites is observed between 2076 and 2090 cm^{-1}. Two phenomena are clearly observed for both the ZrO_2 and TiO_2 based catalysts: after exposure to sulfuric acid decomposition reaction conditions, the wavenumber increases and the intensity of absorption decreases. The

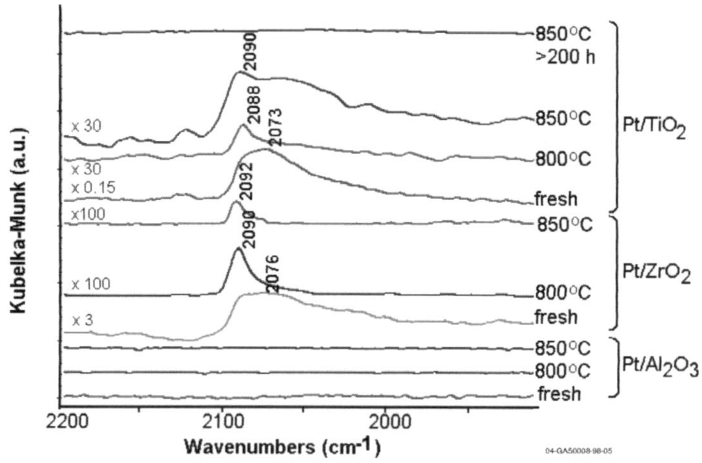

Figure 4. DRIFTS analysis using CO chemisorption.

decreased intensity indicates that less CO is chemisorbed onto the catalysts while the increase in wavenumber points toward a poorer ability of the Pt to donate electrons to the probe molecule. For the TiO_2 catalyst exposed to over 200 h of reaction conditions, no peak is detectable.

One additional catalyst was examined in support of the integrated laboratory-scale efforts being conducted at Sandia National Laboratory (SNL). A 1% Pt/ZrO_2 catalyst, provided by SNL and obtained from Shell Chemical Co. was examined under the conditions listed above. With the higher Pt loading, the catalyst had similar activity to the 0.1 wt % Pt catalyst, however, it appeared to be more stable, as shown in Figure 5. At 800 °C, sulfuric acid conversion was about 40 % during the first day of operation and 35 % on day Two. The reaction was run for a third day; however, the reactor inlet pressure exceeded the relief valve pressure setting of 15 psi between days two and three, thus no data is shown for that last day. Several experiments were run that confirmed that pressure build up at 800 °C. At 850 °C, pressure excursions did not occur for tests up to 73 h, but a slow pressure increase was observed. However, at the end of that test, the catalyst was observed to be physically breaking apart. At this temperature, catalyst activity linearly declined from a first day conversion average of 60% to a fourth day average of 43%. Thus, this 1% Pt/ZrO_2 catalyst displayed plugging problems at 800 °C, and attrition and deactivation problems at both temperatures. Similar to the 0.1 wt% Pt catalyst, BET surface area was found to decrease with exposure to reaction conditions. Surface area declined from a fresh value of 61 m^2/g to 22 and 21 m^2/g following reaction at 800 and 850 °C, respectively. Similar to the titania supported catalyst run for an extended period, no FTIR peaks from CO chemisorption were observed in the DRIFTS analysis for these catalyst samples.

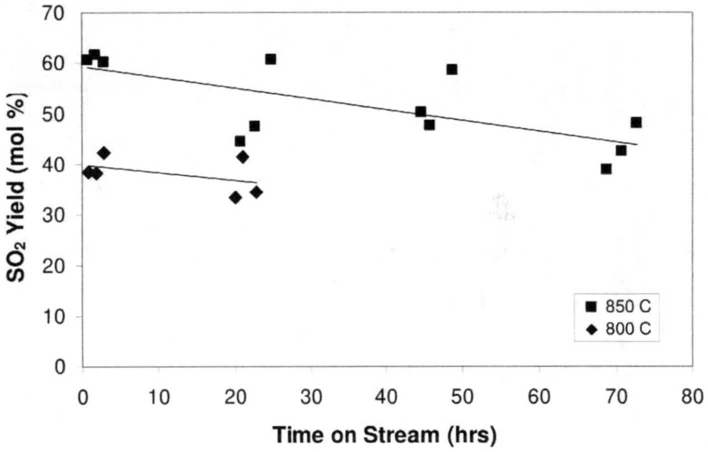

Figure 5. 1wt% Pt/ZrO_2 catalyst stability at 800 and 850 °C.

CONCLUSION

The two higher surface area catalysts demonstrated higher initial catalytic activity as well as deactivation in time scales as short as 6 hours. The lower surface area Pt/TiO_2 displayed lower initial activity but relatively higher stability. The titania supported catalyst provided significant levels of activity for over 200 hours of continuous sulfuric acid decomposition. Exposure to reaction conditions resulted in significant declines in CO chemisorption for all catalysts and decreased support surface area for the alumina and zircona supported catalysts. The sulfuric acid decomposition provides an extremely harsh environment for catalytic materials and stable, highly active catalysts have not yet been identified.

ACKNOWLEDGMENT

This work was supported by the U.S. Department of Energy, Office of Nuclear Energy, Science, and Technology, and the Idaho National Laboratory's Laboratory Directed Research & Development (LDRD) Program under DOE-NE Idaho Operations Office Contract DE AC07 05ID14517.

REFERENCES

1. Brown, L.C., Besenbruch, G.E., Lentsch, R.D., Schultz, K.R., Funk, J.F., Pickard, P.S., Marshall, A.C., Showalter, S.K., General Atomic Report GA-A24285, June 2003.
2. Spewock, S., Brecher, L.E., Talko, F., Proc. 1st World Hydrogen Energy Conf. 1976, Vol 3.
3. Norman, J.H., Mysels, K.J., O'Keefe, D.R., Stowell, S.A., Williamson, D.G., Proc. 2nd World Hydrogen Energy Conf., 1978, Vol 2: p 513.
4. Norman, J.H., Besenbruch, G.E., O'Keefe, D.R., General Atomic Report GA-A15267, March 1981.
5. Norman, J.H., Mysels, K.J., Sharp, R., Williamson, D., Int. J. Hydrogen Energy, 1982, 7, p. 545 556.
6. Tagawa, H., Endo, T., Int. J. Hydrogen Energy, 1989, 14, p.11-17.
7. High-Pressure Catalytic Metal Reactor in a Simulated Solar Central Receiver, GA Technologies Report GA-A18285, February 1986.

76c Life Cycle Assessment of Ispra Mark 9 Thermochemical Cycle for Hydrogen Production

Bradley J Ward, Richard D Boardman, Vivek Utgikar

Various thermochemical water-splitting cycles have been proposed in the literature for producing hydrogen, a potential energy carrier of the future. A complete and thorough comparison of these cycles requires that the environmental impacts of the cycles be taken into account. A life cycle assessment (LCA) of the ISPRA Mark 9 (IM9) thermochemical cycle was conducted to quantify the emissions, resource utilization and the energy consumption of the process. The LCA process involved reviewing the existing literature information on the target cycle, and performing detailed material and energy balances on all the operations required for splitting of water. The aspects analyzed included input energy requirements, process efficiencies, and net energy output. A process simulation and analysis computer code was used throughout. Thermal energy efficiencies were determined as the ratio of the energy of products to the energy of inputs. Overall thermal efficiencies for the IM9 were compared with those for sulfur-iodine (SI) and UT-3 cycles. The maximum temperature in the IM9 cycle is significantly lower than the leading cycles and as such it may offer some advantages related to materials of construction and equipment life over other cycles. The LCA presented in this paper is useful in developing a better understanding of the overall impact of the cycle and providing information that assists in decision-making regarding alternative hydrogen production processes.

Can Sonochemistry Help in Producing Hydrogen Along Thermochemical Cycles?

A. Rahier, SCK•CEN, 200 Boeretang, B-2400, Mol, Belgium

1. Abstract

Sonochemistry is known to favor specific chemical routes, especially those involving radicals. However, the main chemical effects of sonication result from intense cavitation, which is not easily obtained under high pressures. Nevertheless, some difficult steps of known thermochemical cycles may advantageously make use of sonochemistry. After having listed several examples where sonochemistry is likely to improve the unit operations, we will specifically focus on the I-S cycle. More particularly, we will assess the perspective of decomposing HI into H_2 and I_2 without necessarily separating HI from H_2SO_4.

2. Introduction

Among all possible hydrogen production processes that make use of nuclear energy, thermochemical water splitting cycles are very promising [1]. During the last decades, many different cycles have been proposed and were also carefully screened and compared. Several selection criterions were carefully chosen, including the technological feasibility (peak temperature, corrosiveness etc.), the number of reactions and separation steps, the expected energy efficiency, the development progress status as well as economical and safety related issues. Although most of the retained cycles allow to dramatically reducing the peak temperature needed for splitting water efficiently, the latter temperature still remains quite close to the maximum allowable value for state-of-the-art demonstrated designs. Therefore, several attempts are made to further reduce the maximum temperature, either through proposing alternate cycles, or by focusing on the possible enhancement of existing cycles. The present paper examines potentially new perspectives that are possibly offered by sonication in this field. After having recalled some basic characteristics of sonochemistry, we briefly discuss its possible use in the case of two promising cycles. Next we examine the I – S cycle in more details. Very few (essentially qualitative) experimental results will be presented here. Rather, we aim at examining the perspectives offered by sonochemistry, thereby deriving some research lines that could eventually deliver promising results to help in facing the technological challenge of implementing an efficient thermochemical cycle.

3. Basic considerations

There are two incentives to examine the possible application of sonochemistry to water splitting cycles. The first one results from the known fundaments of sonochemistry, which rely on the very high temperatures (> 5000 K) reached just after the implosion of cavitation bubbles. The second incentive is related to the known fact that sonochemistry acts often markedly on reactions involving radicals while such reactions are commonly encountered as part of many proposed thermochemical cycles. But counter arguments exist also according to which sonication might not necessarily provide valuable help to lower process temperatures on an efficient way. Indeed, since it relies on the implosion of cavitation bubbles, sonochemical effects are essentially useful in liquid but not in gaseous phases. Therefore, sonication will not help in the many cases where reactions should inevitably be carried out in the gaseous state, unless alternate paths and/or concepts are found for achieving the same results. It is also known that high pressures and/or the presence of dissolved gases influences the cavitation threshold: under high pressures, both the cavitation threshold and the violence of bubble collapse are increased, while dissolved gases tend to fill up cavitation bubbles, thereby decreasing the local temperature after implosion. These effects must be kept in mind when considering sonochemistry as a potential tool in particular applications.

4. Potentially useful contributions of sonochemistry to promising thermochemical cycles

4.1. Cycle 1: the MARK family of cycles

The so-called MARK 1 cycle was proposed as early as 1969. It relies on the following set of reactions [2]:

$$CaBr_2 + 2\,H_2O \xrightarrow{1000\,K} Ca(OH)_2 + 2\,HBr \quad (1)$$

$$Hg + 2\,HBr \xrightarrow{520\,K} HgBr_2 + H_2 \quad (2)$$

$$HgBr_2 + Ca(OH)_2 \xrightarrow{500\,K} CaBr_2 + HgO + H_2O \quad (3)$$

$$HgO \xrightarrow{870\,K} Hg + \frac{1}{2}O_2 \quad (4)$$

This cycle has been modified several times. One recent version (MARK 13) ends up in a hybrid cycle that combines electrolysis with thermochemistry, as proposed by ISPRA:

$$Br_2 + SO_2 + 2\,H_2O \xrightarrow{390\,K} H_2SO_4 + 2\,HBr \quad (5)$$

$$2\,HBr \xrightarrow{Electrolysis} H_2 + Br_2 \quad (6)$$

$$2\,H_2SO_4 \xrightarrow{1200\,K} 2\,H_2O + 2\,SO_2 + O_2 \quad (7)$$

The MARK 1 approach has been largely criticized because it makes use of mercury, a metal that is not environmentally friendly. In the past, several efforts were made to cope with this but sonochemistry was not considered. Assume that instead of using mercury (reactions 2 – 4), we submit an aqueous solution of HBr to intense sonication. As is known, HBr and HI are likely to undergo reactions involving radicals. Moreover, it is also known that iodide can be oxidized to deliver iodine under intense sonication [3]. Hence, we may expect that reaction 8 may help eliminating the need of mercury:

$$2\,HBr \xrightarrow{)))} H_2 + Br_2 \quad (8)$$

This idea appears not to be easily implemented in the practice. In fact, we lack experimental data to conclude about the viability of the suggested path. We know that demonstrated mechanisms in the case of iodine (equations 9 – 12) would not be applicable to bromine due to the oxidative properties of Br_2 (reaction 12 would not run conveniently with bromide).

$$H_2O \xrightarrow{)))} H^{\circ} + OH^{\circ} \quad (9)$$

$$2\,H^{\circ} \rightarrow H_2 \quad (10)$$

$$2\,OH^{\circ} \rightarrow H_2O_2 \quad (11)$$

$$H_2O_2 + 2\,I^{-} + 2\,H^{+} \rightarrow I_2 + 2\,H_2O \quad (12)$$

But in parallel to the above-mentioned equations, we may also observe reactions 13 – 15:

$$HBr \xrightarrow{)))} H^{\circ} + Br^{\circ} \quad (13)$$

$$2\,H^{\circ} \rightarrow H_2 \quad (14)$$

$$2\,Br^{\circ} \rightarrow Br_2 \quad (15)$$

The viability of such a mechanism requires to efficiently converting sonic energy into chemical energy, specifically for the decomposition of the acid. *A priori*, there is no reason to think that the sonochemical decomposition of HBr would be impossible. At least, the energy needed to dissociate HBr is less than the corresponding energy to dissociate water (reaction 9). Rather, the most acute questions concern the conversion yields that can be expected as well as the overall energy efficiency. Indeed, in the case of iodine (equations 9 – 12), it is known that the energy efficiency is very low [3]. Therefore, sonochemistry should be complemented by classical means to favor efficient energy conversion rather than being used as a unique driving force. It is always advisable to focus on the combination of thermochemical (T, P) and kinetic (suitable catalyst to promote the efficient use of energy) conditions with sonication. Finding out such combinations is not excluded provided appropriate research is conducted in this sense. If successful, the method would allow replacing the MARK 1 set of equations by reactions 16 – 18. In the present case, we think that specific research on this would not be really justified, even regarding the potentially expected results, unless better performances than those associated to the MARK 13 scheme can be reasonably targeted, which appears to be doubtful. As is the case for the MARK 13 cycle, the sonochemical method would eliminate the need to use mercury (or any other metal as ever proposed), but the peak temperature of the cycle would not be lowered. Note that generating sonic energy requires also electrical energy, as is the case for electrolysis.

$$CaBr_2 + 2\,H_2O \xrightarrow{1000\,K} Ca(OH)_2 + 2\,HBr \quad (16)$$

$$2\,HBr \xrightarrow{)))} H_2 + Br_2 \quad (17)$$

$$Ca(OH)_2 + Br_2 \xrightarrow{950\,K} CaBr_2 + H_2O + \frac{1}{2}O_2 \quad (18)$$

4.2. Cycle 2: the UT – 3 scheme

The University of Tokyo has proposed this cycle, which relies on steps 19 – 22:

$$Br_2 + CaO \xrightarrow{950\,K} CaBr_2 + \frac{1}{2}O_2 \quad (19)$$

$$3\,FeBr_2 + 4\,H_2O \xrightarrow{870\,K} Fe_3O_4 + 6\,HBr + H_2 \quad (20)$$

$$CaBr_2 + H_2O \xrightarrow{1000\,K} CaO + 2\,HBr \quad (21)$$

$$Fe_3O_4 + 8\,HBr \xrightarrow{500\,K} Br_2 + 3\,FeBr_2 + 4\,H_2O \quad (22)$$

All these reactions involve solids. Therefore, one can doubt that sonochemistry could be really useful, at least through its recognized chemical effects resulting from cavitation. But besides the fact that ultrasonic waves may accelerate chemical processes through their hydrodynamic effects, one subtler question with this cycle would be: do we really need calcium and iron to carry out the overall decomposition of water? Consider the endothermic reaction 23:

$$Br_2 + H_2O \xrightarrow{)))} 2\,HBr + \frac{1}{2}O_2 \quad (23)$$

The reverse reaction is known to run spontaneously but quite slowly under standard conditions (i.e. at room temperature, in liquid phase, and without sonication). Here, the challenge would be to find a window in the (T, p) space that would:

- Favor sufficiently reaction 23 by increasing the equilibrium constant;
- Allow to keep the main phase in the liquid state;
- Allow checking the effects of sonication on the reaction rate.

We doubt heavily that such (very challenging) conditions exist, but if they exist, we may expect that the proposed approach could ever make sense, certainly if sonication accelerates the reaction rates and if adequate means for extracting the oxygen (e.g. using appropriate membranes as ever suggested) can be used to further favor reaction 23. Note that reaction 23 is not far from being as difficult as the direct decomposition of water from a thermodynamic point of view.

By combining equation 23 with either equation 6 or equation 8, the full cycle would be closed without needing solid materials and hopefully under sufficiently mild conditions.

5. The iodine – sulfur cycle

This cycle is one of the most popular. It has been retained by DOE [1] and by JAERI [4]. The corresponding reactions are given by equations 24 – 26:

$$I_2 + SO_2 + 2\,H_2O \xrightarrow{400\,K} H_2SO_4 + 2\,HI \quad \text{(24; Bunsen reaction)}$$

$$2\,HI \xrightarrow{670\,\text{K}} H_2 + I_2 \quad (25)$$

$$H_2SO_4 \xrightarrow{1150\,\text{K}} SO_2 + H_2O + \frac{1}{2}O_2 \quad (26)$$

The proposed scheme is attractive for several reasons:

- It does not require solid phases and does not involve many reactions and/or separation steps;
- One may expect that adequate materials can be found to withstand the peak temperature, although corrosion problems may be acute;
- It has been studied for a long time and it is well documented.

It has also some drawbacks:

- The chemical products are corrosive;
- The chemistry of $HI – I_2 – H_2SO_4 – H_2O$ mixtures is complicated.

Considering equations 9 – 12, and in a first attempt to assess the need to separate the acids, we carried out a very first purely qualitative test. We submitted a mixture of sulfuric acid (300 mL H_2SO_4 7 M) and hydroiodic acid (100 mL HI 57 wt. %) to sonication (20 kHz, 30 W input power). Since our equipment does not yet allow us to carry out experiments above atmospheric pressure and at high temperatures, we performed the test starting from room temperature. By the time we carried out the test, we had no adequate equipment to detect hydrogen. Iodine was qualitatively detected by extraction in carbon tetrachloride. Within 15 minutes, we observed the following:

- An increase of the temperature to 320 K: this is due to the partial conversion of sonic energy into heat;
- The fast production of huge amounts of iodine (see below for a possible explanation);
- The production of some hydrogen sulfide.

We think that in the chosen conditions, sonication accelerate the parasitic reaction given by equation 27:

$$H_2SO_4 + 8\,HI \xrightarrow{)))} H_2S + 4\,I_2 + 4\,H_2O \quad (27)$$

Although one may think at the recovery of hydrogen from hydrogen sulfide, and although iodine appears to be easily recovered through this reaction, the latter is parasitic because of the following drawbacks:

- Only the hydrogen carried by sulfuric acid could be recovered through the dissociation of hydrogen sulfide, not the hydrogen carried by HI as desired. The latter turns back into water;
- Only 25 % of the sulfuric acid produced by the Bunsen reaction can undergo reaction 27, while all the available iodide is being re-oxidized to yield iodine. Therefore, closing the cycle requires to maintain reaction 26 together with the acid separation step;
- Additional steps would be needed: the decomposition of hydrogen sulfide and the subsequent oxidation of sulfur into sulfur dioxide.

These observations tend to indicate that sonication does not provide valuable help when the acids have not been separated. However, no definitive conclusion can be derived for now because:

- The possible production of hydrogen still has to be assessed. Both hydrogen and hydrogen sulfide have to be quantified in order to reveal to which extent sonication orientates the mechanisms toward the decomposition of HI into H_2 with respect to the decomposition of HI through reaction 27;
- Still other experimental conditions should be tested: it is not excluded that once hydrogen sulfide is present in sufficiently high amounts, the *in situ* decomposition of HI into H_2 and I_2 would not become predominant. In such a case, extracting hydrogen (e.g. using a Pd – Ag membrane) may be considered to complement the process. Carrying tests at other temperatures is also needed prior to conclude definitively.

Further research lines should focus on the possible use of sonication after having separated the acids. We suggest carrying out experimental tests to assess especially the possibly enhanced dissociation of HI starting from the $HI - I_2 - H_2O$ mixture, by combining sonication with a reactive distillation process as proposed by Roth and Knoche [5].

6. Conclusions

One of the main conclusions that can be derived from the present survey is that attempting to convert sonic energy directly into chemical energy is likely to fail. Rather, sonochemistry may eventually help in orientating reaction mechanisms when used in conjunction with classical techniques. Even so, a direct attempt to dissociate HI in the presence of sulfuric acid indicates that sonochemistry may orientate the mechanisms in the wrong direction. Nevertheless, some possible valuable contributions of sonochemistry have been considered. In order to assess their feasibility, many additional research efforts are needed, the thermochemical splitting of water remaining very challenging. Finally, as an answer to the question "Can sonochemistry help in producing hydrogen along thermochemical cycles?", we identified both difficulties and perspectives. The final answer is that further research is required prior to definitively conclude as to the possible contribution of sonochemistry to the efficient water splitting using thermochemical cycles.

7. References

1. L. C. Brown, G. E. Besenbruch, K. R. Schultz, S. K. Showalter, A. C. Marshall, P. S. Pickard, J. F. Funk, "High efficiency generation of hydrogen fuels using thermochemical cycles and nuclear power", Report GA – A24326, General Atomics, March 2002
2. C. Marchetti, "Hydrogen and Energy", Chemical Economy & Engineering Review, 7 – 24 (January 1973)
3. T. J. Mason, J. P. Lorimer, "Applied Sonochemistry: The Uses of Power Ultrasound in Chemistry and Processing", Wiley – VCH (2002) ISBN 3-527-30205-0
4. K. Onuki, S. Kubo, H. Nakajima, S. Higashi, S. Kasahara, S. Ishiyama, H. Okuda, "R&D on Thermochemical I-S Process at JAERI", Proc. Second Information Exchange Meeting on Nuclear Production of Hydrogen, 111 – 117, Argonne (Illinois, USA), 2 – 3 October 2003
5. M. Roth, K. F. J. Knoche, Hydrogen Energy, 14, 545 – 549 (1989)

76e Lab-Scale Catalytic Decomposition of Sulfuric Acid with Scalable Materials

Fred Gelbard, Paul S Pickard

For both the Sulfur-Iodine (SI) and Hybrid-Sulfur (HyS) thermochemical cycles, a concentrated aqueous sulfuric acid solution is vaporized and catalytically decomposed to produce sulfur dioxide, oxygen, and water. Data and operational experience are needed for these two steps, since they are not part of the conventional operation of a commercial sulfuric acid plant. In particular, very little corrosion data are available under flow conditions at the high temperatures and pressures expected for a commercial operation. In this presentation we report on laboratory-scale experiments that were designed as a test bed to provide the needed thermodynamic, reaction-kinetic, and corrosion data regardless of which cycle is used in a pilot plant and subsequently in an actual plant. The apparatus was constructed completely of industrial metals to demonstrate the feasibility of a large scale process. In addition, special emphasis was placed on incorporating real-time instrumentation for diagnostics and process control. The design, thermodynamic analysis, instrumentation, and initial data will be presented.

76f Conceptual Design for a Hybrid Sulfur Thermochemical Process Plant

Maximilian B Gorensek, Melvin R Buckner, William Summers, Zafar H Qureshi

A conceptual design for a Hybrid Sulfur (HyS) cycle was prepared and modeled using Aspen Plus, and material and energy balances computed. Stream data for this flowsheet, normalized to a 1-kmol/hr H2 production rate, are presented.

The HyS process can be thought of as comprising three different sections. The first section includes the electrolyzers and their associated equipment. Next come the sulfuric acid concentration and decomposition operations, which produce water and gaseous mixtures of SO2 and oxygen. Finally, the need to separate a clean O2 product stream from the SO2/O2 mixture coming from sulfuric acid decomposition, and dissolve the remaining SO2 in the anolyte, leads to a third flowsheet section where these tasks are accomplished. Each of these flowsheet sections are described separately and in detail in the report.

The overall net thermal efficiency for the HyS cycle was calculated as 46.5%. The efficiency was based on the higher heating value of the hydrogen product divided by the total thermal energy requirements, including the thermal energy used to generate electricity with a 50% heat-to-electric conversion efficiency. It should be noted that this efficiency is considerably higher than the estimated thermal efficiency for a direct electrolysis plant, which is expected to be only 36-40% (HHV) maximum. Furthermore, the highest reported efficiency for the SI cycle using the same reactor heat input conditions is only 42%. Higher thermal efficiencies, exceeding 50% HHV-basis, are deemed feasible for both the HyS cycle and the SI cycle for increased reactor outlet temperatures and further optimized process flowsheets.

Preliminary estimates for the capital cost and the hydrogen production cost for an Nth-of-a-kind nuclear hydrogen production plant using the HyS cycle were prepared. The nuclear heat source design and cost prepared for a recent DOE-NE NERI study using General Atomics Modular Helium Reactors and the SI thermochemical cycle were used as a cost basis. The cost of hydrogen production for the HyS Cycle nuclear hydrogen production plant was determined to be $1.73 per kilogram at the plant gate for the baseline case. This cost was $2.03 for higher cost ($3500 per m2) electrolyzers and $1.53 per m2 for lower cost ($1025 per m2) electrolyzers. These costs are nearly the same as the hydrogen cost of $1.65 per kilogram reported for the SI nuclear hydrogen production plant.

76g Hydrogen Production by Water Dissociation Using Mixed-Conducting Dense Ceramic Membranes

Stephen E Dorris, Tae H Lee, Utham Balachandran

Water dissociates into oxygen and hydrogen at high temperatures. The problem with exploiting this reaction is that very low concentrations of hydrogen and oxygen are generated even at relatively high temperatures (e.g., only 0.1 and 0.042% for hydrogen and oxygen, respectively, at 1600 C), because the equilibrium constant for this reaction is small. However, significant amounts of hydrogen or oxygen can be generated at moderate temperatures if the equilibrium is shifted toward dissociation by using a mixed-conducting (i.e., conducts both electrons and ions) membrane to remove either oxygen or hydrogen. We have studied hydrogen production via water splitting at moderate temperatures (500-900 C) with novel mixed-conducting membranes. Hydrogen production rates were investigated as a function of temperature, water partial pressure, membrane thickness, and chemical potential gradient across the membranes. The hydrogen production rate increased with increases in temperature, water partial pressure, and oxygen chemical potential gradient and with a decrease in membrane thickness. The maximum hydrogen production rate, approximately 10 cm3(STP)/min-cm2, was measured with a 0.10-mm-thick membrane at 900 C using 50 vol.% water vapor on one side of the membrane and 80% hydrogen (balance helium) on the other side. A hydrogen production rate of approximately 5 cm3(STP)/min-cm2 was measured at 900 C with a more recently developed membrane (thickness ?1 mm). It is expected that the hydrogen production rate can be increased to greater than 20 cm3(STP)/min-cm2 by reducing the thickness of the newly developed membrane. In these experiments, hydrogen was used as a model gas to establish a high oxygen potential gradient; however, another reducing gas, methane, was substituted in one experiment to maintain the high oxygen potential gradient. With methane as the feed gas, hydrogen was produced on the steam side of the membrane, while synthesis gas (a mixture of hydrogen and carbon monoxide) was produced on the methane side. The oxygen produced by water dissociation was transported across the membrane and reacted with methane to form synthesis gas. The synthesis gas can be subjected to a water-gas shift reaction to produce more hydrogen and a stream enriched in CO2 for sequestration. In this talk, we will present test results of hydrogen production by water dissociation using mixed-conducting membranes.

Work supported by U.S. Department of Energy (DOE), Office of Fossil Energy, National Energy Technology Laboratory's Gasification Technologies Program and DOE, Energy Efficiency and Renewable Energy, Office of Hydrogen, Fuel Cells, and Infrastructure Technologies Program, under Contract W-31-109-Eng-38.

77a Composite Pd/porous Stainless Steel Membranes for the Production of Pure Hydrogen and High Pressure CO2 for Sequestration

Erik E Engwall, Federico Guazzone, Ivan P Mardilovich, M. Engin Ayturk, Yi H Ma

As we are entering the hydrogen economy of the 21st Century, the search for clean, economical and efficient ways of producing pure hydrogen has been intensified worldwide. Although from the environmental point of view, electrolysis of water by solar energy will be the most desirable way for the production of hydrogen, steam reforming of natural gas will remain as the major source for hydrogen production in the US in the foreseeable future. We have developed new and unique ways to prepare composite Pd and Pd/alloy porous stainless steel membranes (PSS) with long term high temperature stability that are ideal for a steam reforming membrane reactor. By using a hydrogen permeable membrane reactor, the added advantage is that the green house gas CO2 is at a high pressure and can readily be sequestrated or used for oil recovery.

Composite Pd/porous stainless steel membranes were prepared by electroless plating. SEM and EDS results showed the oxide layer produced by the in-situ controlled oxidation to be effective as an intermetallic diffusion barrier between the elements in the PSS and the Pd layer. The long term thermal stability of the membrane was further improved by applying additional diffusion barrier layers. Experimental data showed that the membranes with enhanced intermetallic diffusion barriers were stable for hundreds and thousands of hours at 500C. In addition, membranes with thickness less than 10 μm were synthesized and permeances with values over 60 m3/m2-h-atm1/2 at 500C were obtained. Over 95% methane conversion and the production of high pressure CO2 suitable for sequestration or enhanced oil recovery could be achieved using our membrane as a membrane reactor for the steam reforming of methane. In addition, membrane areas as high as ~0.1 m2 have been prepared. Preliminary simple economic analysis indicates that the Pd cost for the membrane is insignificant compared to the total cost of the membrane module as long as the membrane thickness is in the range of 5 to 10 μm.

77b High Temperature, High Pressure Pd-Cu based Hydrogen Purification in the Presence of CO, CO2 and H2O

Bret Howard, Brian Morreale, Micheal Ciocco, Osemwengie U Iyoha, Richard P Killmeyer, Robert Enick

Currently, hydrogen is mostly produced from steam reforming of hydrocarbons or from coal gasification. Hydrogen production from coal gasification can provide a transition to the hydrogen economy: as a result of the large volume of produced hydrogen in addition to the large volume of carbon monoxide which can be further directed to a shift reactor to increase the hydrogen yield. Conventional methods of hydrogen production from hydrocarbons, therefore, produce hydrogen in coexistence with other gaseous components such as CO, CO_2, H_2O and various other side reaction products form which hydrogen must be purified.

Metal membranes, such as Pd-based alloys, possess the potential to produce high purity hydrogen from gaseous mixtures. However, the performance of metal membrane separations may be adversely affected by the presence of various gaseous compounds which may be in co-existence with the hydrogen i.e., the presence of CO and H_2O in a H_2 feed stream reduced H_2 permeance through a Pd-Ag membrane by 40% and 70%, respectively [1]. Several research groups - including the NETL - are exploring the viability of palladium alloy membranes (e.g. Pd-Cu) for post-gasifier water-gas shift (WGS) membrane reactors and separators because of the relatively high hydrogen permeance of these alloys [2] and the H_2S-tolerance of the fcc alloy phase [3]. The primary objective of this study was to determine whether the major constituents of the WGS reaction (CO, CO_2 and H_2O) would adversely affect the efficacy of the Pd-Cu membrane in carrying out H_2 purification.

The viability of an 80%Pd-20%Cu to purify hydrogen in the presence of impurities was assessed by conducting diffusion experiments in the presence of CO, CO_2 and H_2O at high-temperature and high-pressure conditions characteristic of gasifier outlet stream, e.g. pressures up to 1.5 MPa and temperatures up to 1273 K. The results indicate that CO and H_2O have a negligible influence on membrane permeance over the temperature and pressure range of study. Furthermore, a significant roughening of the membrane surface in the presence of H_2O was observed. The membranes operated stably for more than 720 hours. SEM images depicting membrane surface after exposure to mixed feed streams and the severe pitting of the membrane in the presence of H_2O and other issues relating to long term operation of these membranes in the presence of these gases will be discussed.

References

1. Hou, K., et al., Journal of Membrane science 214, (2003) 43
2. Howard, B.H., et al., Journal of Membrane Science 241, (2004) 207
3. Morreale, B., et al., Journal of Membrane Science 241, (2004) 219

77c Dense Cermet Membranes for Hydrogen Separation

Brian Morreale, Ling Chen, Richard P Killmeyer, Stephen E Dorris, Tae H Lee, Utham Balachandran

Argonne National Laboratory (ANL) and the National Energy Technology Laboratory (NETL)are developing dense ceramic membranes for separating hydrogen from mixed gases, particularly product streams generated during coal gasification and/or methane reforming. Hydrogen separation with these membranes is nongalvanic, i.e., it does not use electrodes or an external power supply to drive the separation, and hydrogen selectivity is nearly 100% because the membranes contain no interconnected porosity.

Novel cermet (i.e., ceramic-metal composite) membranes have been developed to separate hydrogen from gas mixtures at high temperature and pressure. The hydrogen permeation rate, or flux, has been measured at atmospheric pressure in the temperature range of 500-900 C for cermet membranes with thickness of 20-200 micron. The hydrogen flux varied linearly with the inverse of membrane thickness, and reached approximately19 cm3[STP]/min-cm2 for a membrane with a thickness of approximately 20 micron at 900 C with 100% H2 as the feed gas. This relationship indicates that the hydrogen flux is controlled by bulk diffusion, and that it might be higher for a thinner (less than 22 micron) membrane. Some of the membranes were tested in a gas mixture containing high concentrations of CO and CO2 for times approaching approximately 300 h, and no performance degradation was observed; this suggests that the membrane is chemically stable and may be suitable for long-term operation. Hydrogen flux measurements in H2S-containing atmospheres show that the cermet membranes are stable in gases containing 200 ppm H2S for up to 500 h. The hydrogen flux of cermet membranes was measured at NETL at hydrogen pressures up to 25 atm and temperature up to 900 C. The present status of membrane development at ANL/NETL will be presented in this talk.

*Work supported by U.S. Department of Energy, Office of Fossil Energy, National Energy Technology Laboratory's Gasification Technologies Program, under Contract W-31-109-Eng-38.

77d Sensing Low Concentrations of Hydrogen by Means of Evanescent Wave Absorption in Pd-Coated Optical Fibers

David Monzón-Hernández, Donato Luna-Moreno, Joel Villatoro

In this work we present different sensor schemes for detecting low concentrations of hydrogen. They all are based on the absorption change of the evanescent waves in palladium-coated optical fibers. Multimode or single-mode tapered fibers coated with a thin Pd layer, holey optical fibers filled with palladium particles, and Bragg gratings inscribed in a side-polished fiber and coated with palladium are proposed as hydrogen sensors. All these sensors are suitable for detecting hydrogen concentrations below 10% at normal conditions. The wavelength at which the measurements are carried out are either 850 nm or 1550 nm. We have found that sensors based on Pd-coated single mode tapered fibers are very sensitivity for concentrations around 4%. On the other hand hydrogen sensors based on evanescent waves interaction of Pd-coated fiber Bragg gratings offer dual optical response. All the sensors proposed here are suitable for remote hydrogen sensing since the optical fibers required exhibit low losses at 850 nm or 1550 nm.

Optical fibre sensors for hydrogen leak detection

Alain Trouillet, Emmanuel Marin, Colette Veillas, Henri Gagnaire

*Laboratory of Signal Processing and Instrumentation, UMR CNRS 5516,
Jean Monnet University, Bat F, 10, rue Barrouin, F-42000 Saint-Etienne
e-mail : alain.trouillet@univ-st-etienne.fr
phone +33 4 77 91 58 12; fax +33 4 77 91 57 81
web : www.univ-st-etienne.fr/tsi*

Abstract

Liquid hydrogen has been intensively used in aerospace applications during the past forty years and is of great interest for future automotive applications. Following upon major explosive risks due to the use of hydrogen in air, several studies were carried out in order to develop optical fibre sensors for the detection of hydrogen leakage. This communication is aimed towards both a review of previously developed sensors and prototypes of hydrogen detection systems based on micro-mirrors or Surface Plasmon Resonance (SPR) and the presentation of new sensors based on the use of Fibre Bragg Gratings (FBG), Long Period Gratings (LPG) and in-series LPG. The sensing principle based on palladium-hydrogen interaction is presented as well as the optical fibre probes. Detection parameters such as response time and sensitivity will be discussed versus temperature.

Keywords : optical fibre sensor, hydrogen sensor, fibre bragg gratings, long period gratings

Introduction

Due to the well-known explosive risks of hydrogen systems, hydrogen sensors are of great interest for aerospace applications and future developments in the fields of fuel cells and automotive applications. A great number of hydrogen sensors[1,2] have been demonstrated and merchandised in the past twenty years. Even they were not specific to hydrogen detection, hydrocarbure sensors have often been used. Most of them provided fast and efficient detection at room temperature and pressure, but their principles and conceptions were not adapted for applications where the need of a deported local detection in a severe atmosphere leads to the choice of optical fibre sensors. Among the numerous methods for gas detection two very different principles have been used for hydrogen detection. They are based on spectroscopic analysis via infrared absorption or Raman scattering[3,4] for the first one and interaction with a transducer which is a thin palladium layer[5] in most cases. Several methods have been demonstrated, the most significant were based on surface plasmon resonance[6], intensity variations by means of electrochromatic materials[7,8] or interferometric measurements[9]. Following upon the works of M.A.Butler[10,11] studies have been carried out in order to develop optical fibre sensors for the

detection of hydrogen leakage. The most relevant of them were based on a nanometric palladium micro-mirror deposited on the cleaved out-put end face of a multimode optical fibre[12] or based on surface plasmon resonance at the interface between the bare core of a multimode optical fibre and a deposited palladium nano-layer[13]. All the prototypes and developed sensors were based on the monitoring of intensity variations of a monochromatic light propagated through the optical fibre and the potentialities of the SPR sensors for volumetric detection have also been demonstrated.
Nevertheless, the requirements of the next generation of sensors must also be considered[14]. In addition to improve functionality at lower cost, the new sensors, regardless of operating technology, must be configured in networks providing global information from real time data. Due to their addressing capabilities, the use of Fiber Bragg Gratings (FBG) and Long Period Gratings (LPG) could contribute to the development of such a new sensor generation. LPG have already been used successfully as solution concentration sensors where the cladding modes interacted with the solution into which the sensor was immersed[15,16]. We are presenting optical fibre hydrogen sensors based on FBG and LPG on the surface of which a palladium nano-layer is deposited. Preliminary results obtained in an in-fibre Mach-Zehnder interferometer configuration with in-series LPG sensors will also be presented. They show potential interest to compensate the thermal sensitivity of the fibre gratings.

Detection Principle

All the developed sensors require the optical fibre probe to be coated by a thin palladium layer that will undergo changes in its optical properties when exposed to gaseous hydrogen. Palladium has the property of absorbing up to 900 times its volume in hydrogen at room temperature and pressure leading to the formation of the hydride PdH_x (where x is the atomic ratio H/Pd). The physical properties of the palladium-hydrogen system are strongly dependant on the gaseous hydrogen pressure and temperature. They are characterized by the existence of two different crystallographic phases called α and β separated by a phase transition. Earlier studies[17] have shown that in the α-phase, which also corresponds to the crystallographic phase of pure palladium, hydrogen is only slightly absorbed by palladium. On the opposite in the β-phase, the absorption of hydrogen in palladium is enhanced. The phase transition is characterized by fast changes in the absorption properties of hydrogen in palladium. Following upon the palladium-hydrogen system properties, the response and response time of the sensor are functions of gaseous hydrogen pressure and temperature. Previous studies performed in our laboratory[12] have shown that due to the weak absorption of hydrogen in palladium, a weak response and a short response time are expected in the α-phase. On the opposite and due to the enhancement of the absorption of hydrogen in palladium, a strong and saturated response which is independent on the hydrogen concentration and a very long response time which can exceed one hour depending on the temperature are expected in the β-phase. Furthermore, preliminary studies have also shown that the phase transition has an important influence on the repeatability of the performances. Depending on the final processing of the data obtained from the sensor, it is therefore necessary to ensure operating conditions of the sensor either in the α-phase or in the β-phase to keep repeatable performances and to avoid as much as possible the operation in the phase transition.

Micro-mirror sensors

Micro-mirror sensors[12] were based on the monitoring of intensity variations of a monochromatic light reflected by a nanometric palladium micro-mirror which was thermally evaporated on the cleaved out-put end face of a strongly multimode optical fibre. They show capabilities for the detection of hydrogen concentrations lower or superior than the explosive-limit (4%) in a wide range of temperatures from –196°C to + 300°C. Following upon the properties of the palladium-hydrogen interaction and in order to provide a response time always lower than 10 seconds, it was necessary to ensure operating conditions in the α-phase. The occurence of the phase transition being strongly dependent on the temperature of the palladium-hydrogen system, the principle of an optical heating of the palladium micro-mirror developed in the PhD thesis of Xavier Bévenot[19,20] was adopted. In this method an increase of the optical power of the laser diode allows to shift the phase transition towards lower temperatures and, as a consequence, to reduce the response times.

The prototype optical fibre system described on Figure 1 has been designed and developed for SNECMA Motors[18] for the detection of gaseous hydrogen leakage near the connecting flanges of the liquid hydrogen pipes on the test bench of the engine Vulcain of the european rocket ARIANE V. It is composed of four independent parts : the optical fibre probe consisting of a silica/silica optical fibre with 400 µm core diameter, 0.22 numerical aperture and 50 cm length on the output of which is a 25 nm-thick palladium micro-mirror; a 5 m length optical fibre extension; an optical fibre Y-junction and an optoelectronic device. The first channel of the optical fibre Y-junction is used for the input of the light emitted by a laser diode and the second channel is used for the output collecting the proportion of light reflected by the palladium micro-mirror In this device, the optical source and the detector are aligned with the axis of the optical fibre to which they are connected without any adaptative optics. The sensing part of the optical fibre probe is protected by a 1.1 mm diameter metallic tube and the full optical fibre probe is protected by a polymeric over-coating placed at the end of the fabrication process.

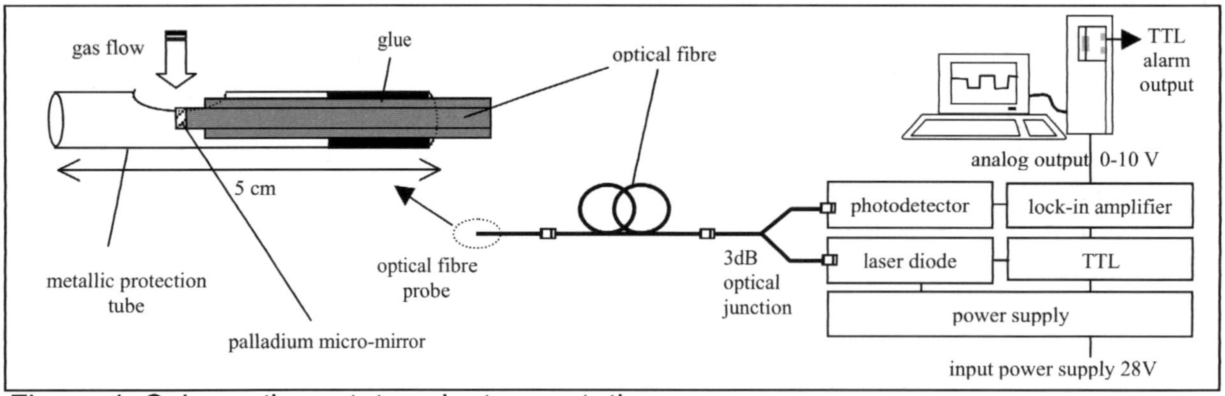

Figure 1: Schematic prototype instrumentation

The typical shape of the detection signal for a 4 % hydrogen concentration is given on Figure 2. The operation of the prototype was characterized by a very low noise (< 0.1%) of the signal provided by the lock-in amplifier. Due to a frequent update of the reference signal, no false alarm activation has been observed with a 0.1% signal variation threshold.

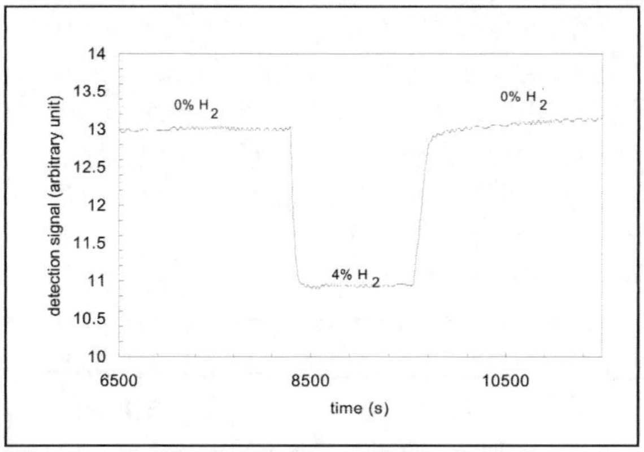

Figure 2: Typical detection signal.

Experimental results with a 15 mW power laser diode have shown that the detection is achieved with a response time lower than 10 seconds for hydrogen concentrations exceeding 0.5% and temperature lower than 100°C and up to a temperature close to 200°C for a 4% concentration. The detection of a 1% concentration hydrogen is also possible for temperatures between 100°C and 140°C but with a degraded reproducibility. It is likewise for a 4% concentration hydrogen leakage and a temperature slightly exceeding 200°C. Furthermore it is also possible to detect hydrogen leakage of concentration lower than 0.5% but only with temperatures lower than 70°C and with response times exceeding 10 seconds. Furthermore, it has been checked the possibility of increasing the accessible temperature range by adapting the laser diode power to a shorter temperature range. For instance, the use of a 1 mW power laser diode allows the detection at temperatures exceeding 300°C with response times between 1 and 3 seconds. A drawback in this case is a response time broadly exceeding 10 seconds at room temperature. On the opposite, the use of a 150 mW power laser diode allows the detection at temperatures lower than -30°C in less than 10 seconds. A drawback in that case is a higher detection limit at a temperature close to only 100°C.

Surface Plasmon Sensors

Surface plasmons resonance (SPR) sensors are well known for their potentialities and applications in the field of chemical and biomedical sensors. First applications with multimode optical fibres have been proposed in the eighties, they were aimed towards the replacement of the bulky Kretschmann prism configuration that could not lead to very simple arrangements because the incident beam must be properly polarized, either collimated or focused. More recently, we have developed a new system, simple and inexpensive, with an optical fibre eliminating both the traditional bulk prism and the spectrometer. This sensor has been completely described previously[21]. It used a multimode optical fibre, which core replaced the high refractive index prism, and a specific injection of a monochromatic light. The core was metallized on the sensing area and a collimated beam was injected into the fiber with the angle of incidence α. It has been shown that some of the skew rays propagated in the fibre could undergo reflections at the core/metal interface with the specific SPR coupling angle θ_{sp}. Thus the light power P transmitted through the entire fibre is attenuated because an amount of the incident optical energy is coupled to the surface plasmon wave.

The experimental set-up for a SPR optical fibre hydrogen sensor is described in Figure 3. The light source is a laser diode emitting a collimated beam at the wavelength 670 nm. The laser diode is attached to a rotation stage and the input end of the fiber has to be positioned on the axis of the precision rotator in order to allow the angle of incidence α to be varied between 0 to 30° with an accuracy of 0.03°. The light transmitted through the fibre is totally collected on a silicon photodiode. This device doesn't need any optical assembly. An optical fibre with numerical aperture higher as possible has been chosen in order to match as close as possible the SPR angle θ_{sp}. It is a multimode HCN-H 400 µm core diameter with numerical aperture 0.48. The fibre is stripped on a short length (20 mm) and the bare core is coated by a thermally evaporated 12 nm palladium layer. The coated sensing part of the optical fibre is dipped into a cell in which a mixture of nitrogen and hydrogen could flow. Measurements of the optical power transmitted through the fibre versus α have been performed with different proportions of hydrogen in pure nitrogen. The incidence angle for an optimal response was found equal to 15 degrees for the detection of 100% of hydrogen and 12 degrees for the detection of 4 % of hydrogen which were the reference values of the study[13]. Figure 4 represents the typical detection signal for successive cycles from a pure nitrogen atmosphere to a mixture of 4 % hydrogen in nitrogen. A good repetition of the response and of the response time has been observed. They were respectively found close to 30% for the response and to 22 seconds for the response time. A recovering time of 50 seconds has been observed, it corresponds to the desorbption of hydrogen from the palladium layer. This time could be considered as the inefficiency period of the sensor between two exposures to hydrogen.

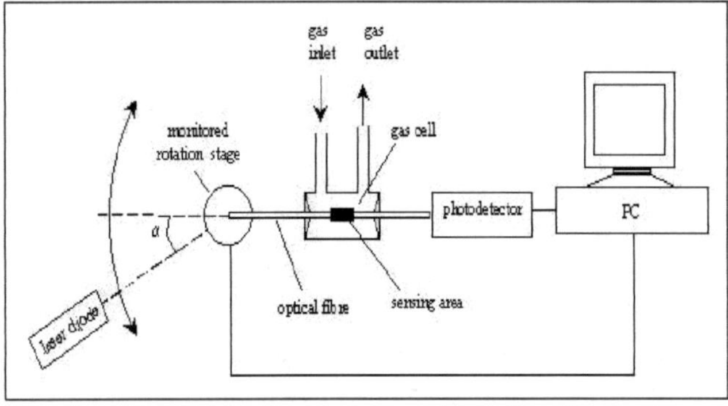

Figure 3 : Experimental set-up for the detection at the temperature +23°C.

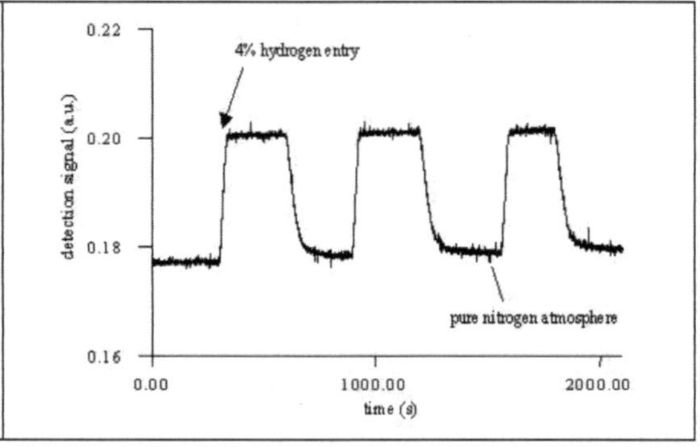

Figure 4 : Typical detection signal for 4% hydrogen and a SPR sensor.

Figures 5a,b show the response and the response time for different hydrogen concentrations in pure nitrogen and compare the results obtained with both the micro-mirror and SPR sensors. Three different behaviour have been observed. For concentrations lower than 1% one can observe only little variations of the response between 1 and 3.5 % corresponding to the α-phase. In the narrow range [1 %, 3 %], one can observe the phase transition and the response strongly varies from 3.5% to 30%. For concentrations higher than 3% palladium is in the β-phase and the 30% response exhibits only little variations with the hydrogen concentration. Concerning the response time, a strong dependance with the concentration has been observed. Very long values of the response time were found for concentrations close to the phase transition.

One can observe on Figure 5a that the shape of the response is identical in the α-phase and that the response of the SPR sensor is two times higher in the β-phase. Figure 5b shows that the evolution of the response time is strictly identical for both sensors. All theses experimental results exhibit a non linearity of the response versus hydrogen concentration. It is therefore impossible to make a quantitative detection of the hydrogen concentration. Nevertheless the analysis of the response time could allow this quantitative detection in the range [1%, 100%]. Furthermore this sensor allows fast and accurate detection for concentrations near and superior to the lower explosive limit (4%). Further improvements of the SPR sensor have shown potential interest in the way of multiple distributed sensors along the fibre for volumetric detection in a closed area for instance.

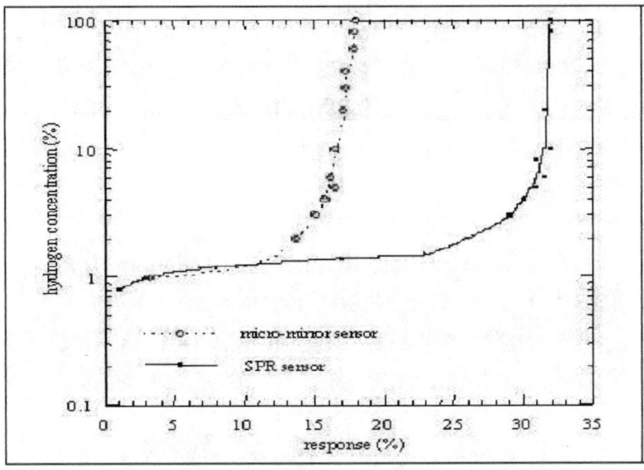

Figure 5a : Response of the sensors

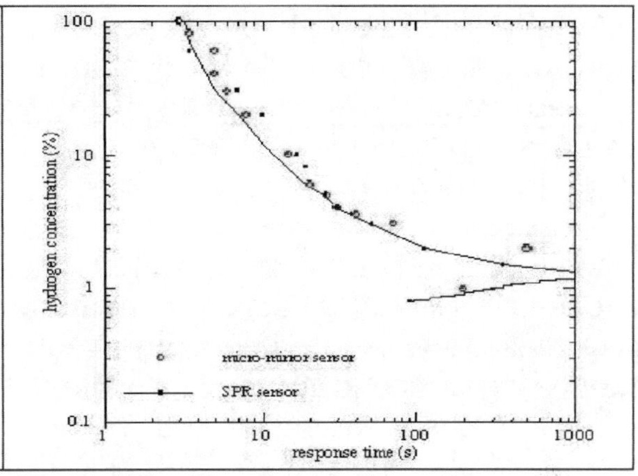

Figure 5b : Response time

In-fibre gratings sensors

Above described sensors present interesting capabilities for local and volumetric detection of hydrogen but do not allow for networks of sensors or applications in which addressing is required. In-fibre gratings could be an answer to such a problem. Fibre Bragg Gratings (FBG) and Long Period Gratings (LPG) are intrinsic devices that allow for the control of the properties of the light propagated inside the fibre. They generally consist of a periodic modulation of the refractive index of the optical fibre core. The presence of the fibre grating promotes the coupling of the fundamental guided mode in the core with forward or backward modes. The coupling is accomplished under the phase matching condition :

$$\lambda = [n_{eff}^{o}(\lambda) - n_{eff}^{i}(\lambda)].\Lambda$$

where $n_{eff}^{o}(\lambda)$ is the effective refractive index of the core mode, $n_{eff}^{i}(\lambda)$ the effective refractive index of the i-th propagated modes and Λ is the grating period. In a FBG the period is in the same order as the optical wavelength and it follows that only coupling to the backward propagated fundamental mode is allowed. The FBG behaves like a wavelength selective mirror at the phase matching wavelength. In a LPG the period could range from 0.1 mm to about 1 mm and it follows that only coupling to forward propagated modes in the cladding is allowed. Because the cladding modes are rapidly attenuated a typical transmission spectrum of such a fibre will show a number of attenuation bands. If the grating as a lentgh L and the coupling coefficient for the i-th cladding mode is κ_i, the minimum transmission of the attenuation bands are found from[15] :

$$T_i = 1 - \sin^2(\kappa_i.L)$$

Changes in the parameters pertaining to the environment of the fibre grating such as temperature, strain and refractive index of the medium surrounding the fibre cladding could change the grating period and the effective refractive indices, resulting in changes in the phase matching condition that will modify the attenuation bands and their minimum transmission values. Connecting variations to the strain, temperature and environmental refractive index variations supported by the optical fibre is nevertheless quite complex since the variations result from a multiple variables system which inversion do not generally admit a single solution. That's why most of the FBG and LPG sensors need a thermal correction that could be made by using a second FBG or LPG sensor which do not support the parameter of influence.

FBG and LPG were written into SMF 28 optical fibres of which the photosensitivity was enhanced by hydrogen loading. A UV laser was used with a phase mask to write short period gratings into the fibre's core after removal of the jacket. Long period gratings were written either by using arc discharges or a UV laser light focused through a 100 mm focal cylindrical lens. The gratings were in the order of 20 mm long and the period was 0.5 µm for the FBG and 400, 420 or 680 µm for the LPG. The resulting FBG had an attenuation band at a resonant wavelength in the order of 1550 nm and each LPG had several attenuation bands at resonant wavelengths varying from 1300 to 1600 nm. Having completed the fabrication of the FBG and LPG, the palladium coating is thermally evaporated to the region where the jacket has been removed previously. Layer thickness of 27, 55 and 62 nm were realized. The experimental set-up to measure the sensitivity of the sensors is described in Figure 6.

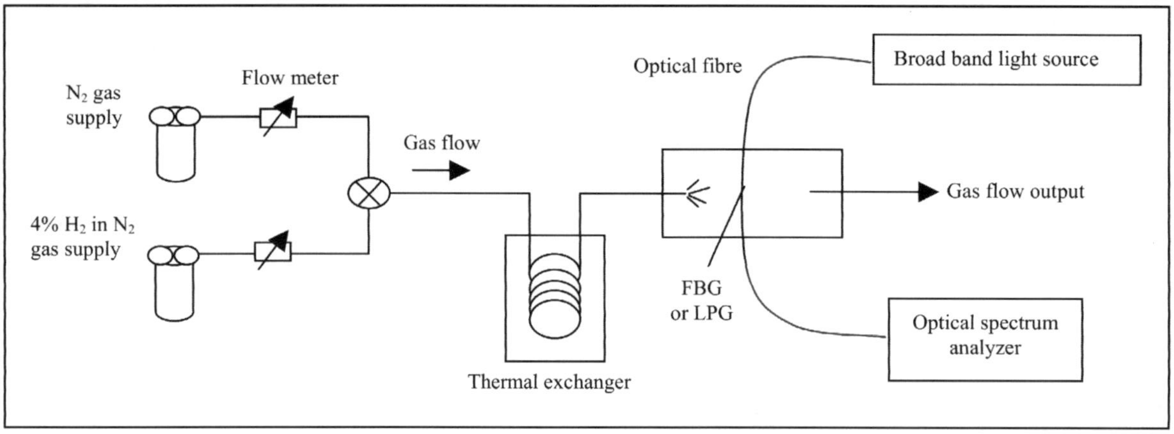

Figure 6 : Experimental set-up

It comprises a set of flow meters and a thermal exchanger in which either pure nitrogen or a preset mixture of 4% hydrogen in pure nitrogen is injected with a gas flow fixed to 500 Nl/h. The thermally regulated gas is delivered to a gas flow tube into which the FBG or LPG sensor is housed. The optical fibre is illuminated by a broadband light source and the experimental date are recorded using an optical spectrum analyzer. The attenuation bands are observed for various temperatures between 0°C and 40°C. The response time of the optical spectrum analyzer being too long to perform response time measurements for the sensor and such a parameter being completely dependant on the palladium hydrogen system that was previously studied, it was set a 2 minutes exposure time to hydrogen before each record. Such a time was chosen to be superior to the response time at any studied temperature.

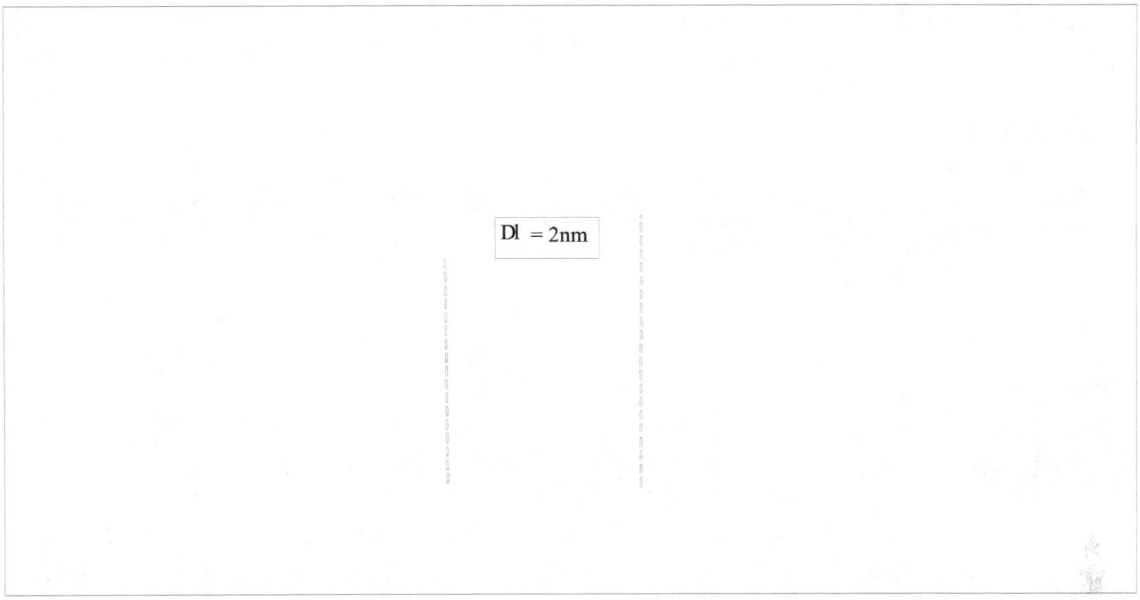

Figure 7 : Attenutation bands of a 680 nm period LPG under exposure to pure nitrogen and 4% hydrogen mixture

Considering first the FBG sensors, in the absence of cladding modes and the fundamental mode being strongly confined into the fibre core no major influence on the refractive effective indices was expected due to the palladium layer. Nevertheless a wavelength shift in the order of 14 pm towards the upper wavelengths was observed at room temperature when the FBG was exposed to the 4% hydrogen mixture. This small shift could be explained by the bending strain following upon the deformation (growing) of the palladium layer when absorbing hydrogen. Considering now the LPG sensors, the presence of cladding modes lead to expect refractive indices variations due to the palladium layer and furthermore coupling between cladding modes and the surface plasmon wave at the palladium layer outer interface. Preliminary results (Figure 7) have shown that a wavelength shift in the order of 2 nm (depending on the grating period) towards the lower wavelengths was obtained in that case. From experiments performed at various temperatures the thermal sensitivity was found independent of the presence of hydrogen in the nitrogen flow.

Further experimental results including the sensitivity to different hydrogen concentrations and results about an in-fibre Mach-Zehnder interferometer using in-series LPG will be presented at the conference. Such a device is expected to show interesting potentialities towards the temperature dependence of the LPG sensors.

Conclusion

The optical fibre sensors presented in this communication show various capabilities to either local or volumetric hydrogen leak detection in a wide range of temperatures. Depending on the configuration, short response times lower than a few seconds could be achieved. Micro-mirror and SPR sensors have also shown capabilities to achieve the detection for concentrations below or superior to the explosive limit. FBG and LPG sensors present further potentialities for future networks of sensors. Very recent results about such in-fibre gratings sensors will be presented at the conference.

Acknowledgements

The authors acknowledge Dr Abdelrafik MALKI from the Laboratory of Electronics, Microtechnology and Instrumentation of the University of Rouen, France who gave LPG fabricated by arc discharge.

References

1. G.W. Hunter, *A survey and analysis of experimental available hydrogen sensors*, Nasa Technical Memorandum 105878, 1992.
2. H.C. Barnes and B.B. Makel, *Quantitative leak detection using microelectronic hydrogen sensors*, 31st AIAA/ASME/SAE/ASEE Joint propulsion conference and exhibit, San Diego, CA, 10-12 July, 1995
3. M.S. Alder-Golden et al., *Laser Raman sensor for measurement of trace-hydrogen gas*, Applied. Optics, 31, pp 831-835, 1992
4. W.A. de Groot, *Fiber-optics based compact gas leak detection system*, 31st AIAA/ASME/SAE/ASEE Joint propulsion conference and exhibit, San Diego, CA, 10-12 July, 1995
5. K. Wyrzykowski et al., *Optical transmission and reflection of PdHx thin films*, J. of Condens. Matter, 1, pp 2269-2277, 1989
6. B. Chadwick et al., *A hydrogen sensor based on the optical generation of surface plasmons in a palladium alloy*, Sensors and Actuators B, 17, pp 215-220, 1994
7. K. Ito and T. Ohgami, *Hydrogen detection based on coloration of anodic tungsten oxide film*, Appl. Phys. Lett., 60, pp 938-940, 1992
8. J. Hamagami et al., *Preparation and characterization of an optically detectable H2 gas sensor consisting of Pd/MoO3 thin films*, Sensors and Actuators B, 13-14 (1993) 281-283.
9. A. Bearzotti et al., *Integrated optic sensor for the detection of H2 concentrations*, Sensors and Actuators B, 7, pp 685-688, 1992
10. M.A. Butler, *Optical fiber hydrogen sensor*, Appl. Phys. Lett., 45, pp 1007-1008, 1984
11. M.A. Butler, *Fiber optic sensor for hydrogen concentration near the explosive limit*, J. Elec. Soc., 38, pp 46-47, 1991
12. X. Bévenot et al., *Hydrogen leak detection system using an optical fibre sensor for aerospace applications*. Sensors and Actuators B, 67, 1-3, pp 57-67, 2000

13. X. Bévenot et al., *Surface plasmon hydrogene sensor using an optical fibre*. Measurement Science and Technology, 13, pp 118-124, 2002
14. A.G. Mignani, *Toughts for the future for chemical sensing*, Second European Workshop on Optical Fibre Sensors, Santander, Spain, 9-11 June 2004, SPIE Proceedings, Vol 5502, p 59, 2004
15. S.W. James et al., *Optical fibre long period grating sensors : characteristics and application*, Measurement Science and Technology, 14, pp R49-R61, 2003
16. R. Falciai et al., *Long period gratings as solution concentration sensors*, Sensors and Actuators B, 74, pp 74-77, 2001
17. F.A. Lewis, *The palladium hydrogen system*, Academic Press, New-York, 1967.
18. A. Trouillet et al., *Gaseous hydrogen leakage optical fibre detection system*, Second European Workshop on Optical Fibre Sensors, Santander, Spain, 9-11 June 2004, SPIE Proceedings Vol 5502, pp 92-95, 2004
19. X. Bévenot, PhD thesis, University of Saint-Etienne, 1998
20. X. Bévenot, M. Clément, H. Gagnaire, EU and USA Patent No 98403319.1-2204.
21. A. Trouillet et al., *Chemical sensing by surface plasmon resonance in a multimode optical fibre*, *Pure and Applied Optics*, 5, 227-237, 1996

77f The Experimental Study and Numerical Simulation of Pd-Alloy Hydrogen Sensor

Erik McCullen, Golam Newaz, Gregory W Auner, Lajos Rimai, Linfeng Zhang, Md H Rahman, Ratna Naik, Ron Baird, Simon Ng

A metal-insulator-semiconductor (MIS) type sensor, utilizing a Pd-Cr or Pd-Ni /AlN/n-Si(111) structure are being developed to measure hydrogen concentration. This MIS type sensor works as a capacitor, the Capacitance-Voltage (CV) curve would shift in the presence of hydrogen. At a constant capacitance, the voltage shift is a function of hydrogen concentration. The alloy gates were deposited in a magnetron sputtering chamber and the AlN was grown using Plasma Source Molecular Beam Epitaxy (PSMBE). Adding Cr and Ni would suppress the phase transition of palladium and provide a more stable film, as revealed by Atomic Force Microscopy (AFM). The alloy gates not only extended the dynamic range of the response to hydrogen concentrations from 100 ppm to 50,000 ppm, but also showed a faster turn on/off response time. Moreover, the device with Pd-Cr gate has a negligible baseline and turn-on response drift and is much stable than that from Pd-Ni device. Through X-ray Photoelectron Spectrometer (XPS), surface segregations of Cr and Ni were found. Furthermore, in Pd-Cr alloy, Cr was oxidized on the outer surface while Pd was oxidized in the metal/insulator interface. This may due to the residual oxygen in the magnetron chamber. However, no similar result was shown in the Pd-Ni alloy film. Through discharge ~ time relationship in "stripping" method in electrochemistry, the diffusion coefficient of hydrogen atom in alloy film decreases with the increasing of Cr concentration. Compared with the sensor's response time, this coefficient is still very high. Considering hydrogen absorption/dissociation on Pd alloy surface and diffusion in the thin film, a numerical model was constructed to simulate the sensor's response; the operating parameters of flow rate, temperature effect could be optimized to obtain big response to hydrogen.

77g Pd/Anodically Oxidized TiO2 Sensor Suitable for Detection of H2 Leakage in Various Environments

Yasuhiro Shimizu

TiO_2 thin films were prepared by anodic oxidation of a Ti plate in a H_2SO_4 solution. The TiO_2 films prepared were a mixture of anatase and rutile phase and contained micron-size pores. A Pd electrode was deposited by evaporation on the TiO_2 film surface to make an electrode configuration of Pd (top electrode)/TiO_2 film/Ti (bottom electrode). The sensor thus fabricated showed nonlinear I-V characteristics in both air and N2, but almost ohmic behavior in H2 balanced with air or N_2. The sensor exhibited a fast response to H_2 and fast recovery at 250 degree C in air as well as in N_2 atmosphere. The H2 response of the sensor in air at 250 degree C was found to vary significantly with the pretreatment conditions: additional pretreatment at 600 degree C for 1 h in N_2 resulted in a deteriorated H2 response especially in air, in comparison with that just after the firing at 600 degree C for 1 h in air for ensuring the electrical contact. Thus, the surface oxidation state and morphology of the Pd electrode was suggested to affect the H_2 response of the sensors. It was revealed that the sensor without the additional treatment in N_2 showed superior H_2 sensing performance to the N_2-treated sensor when operated in N_2 atmosphere. In addition, the response to N_2-diluted H_2 was larger than that to air-diluted H_2, and the interference from water vapor was negligible in N2 atmosphere. The influence of the pretreatment conditions on the H_2 response properties could be weakened by a special surface treatment of the Pd electrode, while maintaining fast response and recovery behavior with large magnitude of H_2 response. In addition, the treatment enables us to lower the operating temperature of the sensor down to 100 degree C. The detail of the treatment will be presented at the meeting and possible mechanism will be discussed.

77h Hydrogen Storage as Solidified Ammonia - Breaking the 2015 DOE Barrier

Claus H Christensen, Jens K Norskov, Rasmus Z Sorensen, Tue Johannessen, Ulrich J Quaade

A new hydrogen system using a solid storage medium involving ammonia will be explored.